Mathtastic Foundation Numbers 1 to 6 Teaching Book

Tracy Ashbridge

MEd, Grad Cert, PG Dip, BEd (Hons)

www.Mathtastic.com.au

Mathtastic

Learning Number Step by Step

Tracy Ashbridge

First Edition

2025 Brisbane

© Copyright 2025 Mathtastic: Tracy Ashbridge. All rights reserved

ISBN 978-0-6455822-3-9

The photocopiable pages, number problems booklet and word problems are available at: **https://mathtastic.com.au/foundation-level/**
Password: **oneTOsix**

Contents

Mathtastic 1
Foundation Numbers 1
1 to 6 1
Teaching Book 1
An Introduction to Mathtastic Foundation 6
Teaching a lesson 7
Module 1 – Numbers 1-3 9
   Ideas for teacher lesson (1 hour) 9
Module 2 – What is addition? – the concept of combining 2 groups 13
   Ideas for teacher lesson (1 hour) 13
Module 3 – What is subtraction? – the concept of removing items from a group. 17
   Ideas for teacher lesson (1 hour) 17
Module 4 – Numbers 4-6 20
   Ideas for teacher lesson (1 hour) 20
Module 5 – Addition – concrete to recording. 23
   Ideas for teacher lesson (1 hour) 23
Module 6 – Subtraction – concrete to recording. 26
Ideas for teacher lesson (1 hour) 26
Module 7 – Doubling concept. 29
   Ideas for teacher lesson (1 hour) 29
Module 8 – Sharing concept. 34
   Ideas for teacher lesson (1 hour) 34
Foundation 37
Numbers to 1-6 Resources 37
   Resources – not included 38
      Lesson Plan 39
      Homework - 15 mins per day 41
   Resource 1 Net for cube to make 1-3 dice (or use blank sided dice available from craft sections of discount stores) 43

Resource 2 Number cards – coloured 44

Resource 3 Number cards – no colour 49

Resource 4 Tracing Numbers 54

Resource 5 Number formation reference chart using the hand 56

Resource 6 Memory Game 0-3 dice patterns 57

Resource 7 Number Track Game 61

Resource 8 Five Frames – 5 wise and random 62

Resource 9 Addition and Subtraction actions 64

Resource 10 Cover Up Game 65

Resource 11 Lady Beetle Spots Game (laminate or place into a plastic sleeve) 66

Resource 12 Cover up number formation game cards – cut into 4 playing cards (laminate or place into a plastic sleeve) 67

Resource 13 Snakes and Ladders 68

Resource 14 Finger subitizing 69

Resource 15 Addition prompt page 71

Resource 16 Go Fish addition to 6 72

Resource 17 Addition Forwards Backwards Game 76

Resource 18 Subtraction Prompt Page 77

Resource 19 Subtraction Forwards Backwards Game 78

Resource 20 Subtraction Go Fish or Memory Game 79

Resource 21 Doubles Bus 85

Resource 22 Forwards backwards game doubles 86

Resource 23 Skittles game (laminate or place in plastic sleeve) 87

Resource 24 Teddy Bears Picnic (Print on A3, laminate or place in plastic sleeve) 88

Resource 25 Sharing Game 89

Foundation 90

Numbers to 1-6 Workbook 90

Instructions 91

Setting out the student book 93

Coding the answers 95

Module 5 - Addition 96

Module 6 - Subtraction 96

Module 7 - Doubling 96

Module 8 – Sharing 97

Foundation 98

Numbers to 1-6 98

Word Problem Solving book 98

Recording Page 100

Module 2 – Part part whole - numbers 1-3 101

Module 2 – Addition by joining - numbers 1-3 110

Module 3 - Remove/ subtract - numbers 1-3 120

Module 4 - Addition or Subtraction - numbers 1-3 130

Module 5 - Addition part part whole - numbers 1-6 142

Module 5 - Addition by joining – numbers 1-6 157

Module 6 - Subtraction - numbers 1-6 172

Module 7 - Doubling - numbers 1-6 187

Module 8 - Sharing - numbers 1-6 202

© Copyright 2025 Mathtastic: Tracy Ashbridge. All rights reserved

An Introduction to Mathtastic Foundation

Mathtastic can be used to teach 1:1 or small groups. The Foundation program is a little different to the core programs as it provides activities to ensure the foundation skills are in place. There is one module for each of the 8 areas. This can be taught as a whole lesson or in parts as required to match student needs.

Links to related resources are shown as R1. The resources can be downloaded for easy printing from the website. The number and word problems workbooks can also be downlaoded for you to record and annotate as required.

Video tutorials on the website are shown LF M1 V1 details of the website page and password are on page 2.

The focus areas in the Foundation level book are:

1. Numbers 1-3
2. What is addition?
3. What is subtraction?
4. Numbers 4-6
5. Addition
6. Subtraction
7. Doubling
8. Sharing

The main program (Levels 1 and above) spirals through different levels of numbers, each level addressing the 8 number sense strategies:

1. Add 0,1,2, 3, Subtract 0,1,2, 3
2. Add from largest number by counting on, Subtract by counting back
3. Rainbow facts
4. Adding tens, subtract tens
5. Doubles/ halving
6. Near doubles
7. Partitioning numbers by place value
8. Adding and subtracting by compensating, Bridge 10 with a 9, Bridge 10 with 7 or 8, Round and adjust

Teaching a lesson

There are 8 sections to each lesson.

1. Thinking problems – these are designed to be open ended and challenge the student to think mathematically. This is an opportunity for mathematical discussions and exploration.
2. Subitizing (sub rhymes with cube) – this is the skills of recognising a set of objects without counting and is a key skill which is not always established in students with difficulties in maths.
3. Counting patterns and objects – students need to develop a sense of the number line. This is a skill that students with difficulties are often weak in.
4. Number sense – each session there is a different focus working through the 8 areas of number sense. These are explained and modelled before applying the number sense concept to problems.
5. Game – many students with math difficulties can get anxious about maths and practising the skills through games is a less threatening way to gain the repetition they need. The games have been chosen to specifically practice the skill in focus and also allow for reasoning skills.
6. Word problems – students need to apply their knowledge in problems. These are organised by the 11 different ways of presenting addition and subtraction problems so students don't just learn to solve for the final answer but can be flexible to work around the problem. In the Foundation Level only addition: part part whole and joining are taught. For subtraction only separate result unknown is taught,
7. Number problems – each session there are number problems related to the focus area and for modules 7 and 8 spaced retrieval of focus areas is included.
8. More games

Each module can be used as a lesson or can be split over several lessons depending on the time you have available and the speed the student works through the number sense strategies. The games can be repeated easily, and many have options to extend them.

A blank lesson plan is included in the resources for you to plan for specific students. Homework can be easily set and there is a grid for this. Homework could be games or extra practice from the workbooks.

Module 1 – Numbers 1-3

Ideas for teacher lesson (1 hour)

Thinking Problems	**Thinking task** https://tracyashbridge.com/developing-sorting-skills/ LF M1 V1  Same or different How are 2 groups the same? How are they different? Ask students to sort coloured and different shaped items into groups – e.g. use the 3 sized sorting bears - sort by colour, sort by size, use tangram shapes – sort by colour, sort by shape, category sorting – fruits, animals etc Explore: What do you notice? What do you wonder? 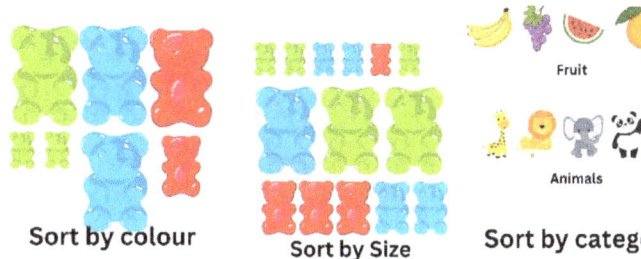
Subitizing 	**Subitizing** R1 or a large blank 6 sided dice. With the student make a 1-3 dice. Teach the dice patterns 1-3 by drawing on the dots onto the dice. Then roll a 1-3 dice and recognise the pattern quickly without counting.

Counting – patterns and objects

Number songs LF M1 V2
- Five little speckled frogs
- Five green bottles
- Five in the bed
- Five Little Monkeys
- Five currant buns
- One, Two, Three, Four, Five Once I Caught a Fish Alive
- Hickory Dickory Dock – add own verses
- One Two Buckle my Shoe

<u>ABAB patterns</u> – copy, continue, create – use concrete materials to build, then move to stickers/ stampers/ drawing. You can use plates or boxes to support the understanding of the repeated pattern.

Do many repeats of the same pattern but with different options e.g. red, blue, red, blue change to yellow, green, yellow, green change to cat, dog, cat dog, change to car, bike, car, bike etc before you change to another pattern structure.

<u>Count to 10</u> R2 or R3 LF M1 V3
Put out the numbers in a rainbow arc and match the pictures and or words – colour coded version available to support – colours to match Cuisenaire rods

 Number Sense	**Number sense focus area explain, explicit teach and model** **Examples and nonexamples** <u>Count</u> objects to 3, pictures to 3 (use magazines and count small collections within pictures) and say the number LF M1 V4  Teach student how to count a group of objects: student to line up objects to count and move them/ touch them as counted, try moving from one colour paper to another. <u>R4</u> & <u>R5</u> LF M1 V5  Trace and write the numbers 0, 1,2,3 – use hand to support correct formation.
 Games	**Game/ hands on activity** <u>R6</u> LF M1 V6 Memory – match number and/or dice pattern 0-3. The student can match numbers and dice pattern as well. 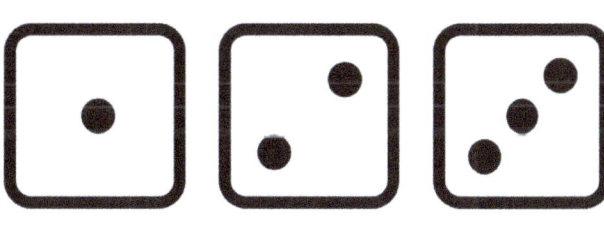  A student could match 2 dice patterns, 2 number digits or one of each.

Word Problems	**Word Problems – Problem Solving Book Foundation Level** – These start at Module 2.
Number Problems	**A: Number problems** – solve the problem in different ways. These start from Module 5.
Games	***Game/ hands on activity*** R7 LF M1 V7 Number Track Game Use the blank number track 0-6. Roll the 3-sided dice and move the correct number along the track. The winner is the first to get to the end of the track. Move along track from left to right the same as if the student is reading. Extension 1. Use a number track 0-6 instead of a blank track. 2. Colour half of the dice red and half green. If the dice is red move to the right and green move to the left. The winner is the first to get to either end of the track.

Module 2 – What is addition? – the concept of combining 2 groups.
Ideas for teacher lesson (1 hour)

Thinking Problems	**Thinking task** LF M2 V1 Combining 2 groups- the envelope game. You will need 7 envelopes or similar marked up 0,1,2,3,4,5,6. Use R2 or R3 but only cards 0,1,2,3. The student picks any 2 cards. Make the cards up with concrete materials, find the total. Place both of the 2 cards into the envelope marked with the answer. Conclude the activity by considering the contents of each envelope: What do you notice? What do you wonder?
Subitizing	**Subitizing** R8 5 frames - just use the cards with 0-3 turtles You may need to teach 0 as a concept – none = zero Take time to look at each card and count up the turtles. Once the student is familiar with these explore which they can subitize. How fast can they do them all? Can they beat their time?
Counting	**Counting – patterns and objects** https://tracyashbridge.com/developing-patterning-skills/ AAB and ABB patterns– copy, continue, create – use concrete materials to build, then move to stickers/ stampers/ drawing Do many repeats of the same pattern but with different options e.g. red, blue, red, blue change to yellow, green, yellow, green change to cat, dog, cat dog, change to car, bike, car, bike etc

Counting continued	R2 or R3 LF M2 V2  Count to 10 – put out number rainbow and matching pictures, include number words - plain or colour coded to link to Cuisenaire rods. R4 LF M2 V3  Trace and write numbers 0 to 3 in multiple ways and then copy onto a number track R5 LF M2 V4  Use hand model to support formation development. LF M2 V5  Count up to 3 objects with 1 to 1 correspondence – check if student understands conservation if numbers - that if you move the items the count is still the same.
 Number Sense	**Number sense focus area explain and model Examples and nonexamples** R9 LF M2 V6 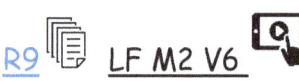 Model counting 2 groups and then pushing them together to make a bigger group – this is called adding. Model some in each hand and cross arms to make add sign.
 Games	**Game/ hands on activity** R1 LF M2 V7  Create a 0-3 dice, choose to double 2 of the numbers.

Games continued	Roll the dice and collect concrete materials to match the dice, roll a second time and repeat. Count up how many altogether. The second player does the same. Compare the 2 results who has most? Least? R10 LF M2 V8 *Extension – cover the number on the board. The player can cover one of the numbers rolled or the total. For example, if you rolled 2 and 3. You could cover 2 or 3 or 5 (the total of both dice). LF M2 V9 *Further extension roll 2 dice and add together. Cover the correct number on the board. The winner is the first to cover all numbers.
Word Problems	**Word Problems – Problem Solving Book Foundation Level** – These start at Module 2. Problem solving tasks will need to be supported with concrete and representative materials – e.g. counters and drawing pictures. Read the question to the student, possibly more than once. What does the question ask you to do? How can you solve this?
Number Problems	**A: Number problems** – solve the problem in different ways. These start from Module 5.

	Game/ hands on activity
 Games	[R11] Lady bug addition, [R1] Create a dice with numbers 0,1,2 LF M2 V10 Roll the dice, draw the dots on one side of the ladybug. Roll again to complete the second side. Add the total together. On a second ladybug sheet, the second player can do the same. The winner is the player with the most or least – players to decide the rules.

Module 3 – What is subtraction? – the concept of removing items from a group.
Ideas for teacher lesson (1 hour)

	Thinking task LF M3 V1 Removing items from the group- the envelope game. Use R2 or R3 but only cards 0-6 You will also need 7 envelopes marked up 0,1,2,3,4,5,6 – use the number cards with picture supports for the first few rounds then just the plain number cards. The student picks any 2 cards. Find the card with the largest number then subtract the card with the lowest number. Place both of the 2 cards into the envelope of the answer. Conclude the activity by considering the contents of each envelope :What do you notice? What do you wonder?
	Subitizing LF M3 V2 Use Dominoes with 0-3 dots in total. Place these in a bag. Pull a domino out of a bag. How many dots can you see? How do you know?
	Counting – patterns and objects abcabc pattern – copy, continue, create – use concrete materials to build, then move to stickers/ stampers/ drawing Do many repeats of the same pattern but with different options e.g. red, blue, red, blue change to yellow, green, yellow, green change to cat, dog, cat dog, change to car, bike, car, bike etc

Counting continued	Use the plate idea from Module 1 to show the repeating pattern  R3 LF M3 V3 Count to 10 – put out number rainbow and matching pictures, include number words - colour coded, start to turn over some of the cards but leave them there to count through R4 & R5 Trace and write numbers to 0-3 using the hand model for support. R1 & R12 LF M3 V4 Roll a dice numbered 0-4 and cover the numbers on the board – winner is first to cover all.
 Number Sense	**Number sense focus area explain and model Examples and nonexamples** Teach concept of subtraction = taking away R9 LF M3 V5 Model some in each hand and cross arms remove an arm to leave take away sign Model with multiple concrete items toys, food (eating them is a good example!), cross out pictures

Games	**Game/ hands on activity** R13 LF M3 V6 Snakes and Ladders game – move counter to match dice count – use dice with digits or dots 1,2 and 3 only - R1
Word Problems	**Word Problems – Problem Solving Book Foundation Level** – Problem solving tasks will need to be supported with concrete and representative materials – e.g. counters and drawing pictures. Read the question to the student, possibly more than once. What does the question ask you to do? How can you solve this?
Number Problems	**A: Number problems** – solve the problem in different ways. These start from Module 5.
Games	**Game/ hands on activity** R1 & R11 LF M3 V7 Lady Beetle spots game – draw or place 3 dots on the lady beetle. Play solo or against a partner. Subtract the number rolled on the dice. Use a 0-3 dice and roll the dice. The winner for each round is the player who has the least number of dots left.

Module 4 – Numbers 4-6
Ideas for teacher lesson (1 hour)

Thinking Problems	**Thinking task** **How Many Are Hiding? LF M4 V1** Place 6 items on the table. Ask the student to look away while you hide some under a cup or something similar. Ask the student to count how many they can see, and then work out how many they can't see. Link this to the idea of 2+?=6 where there is a missing number in the equation. Start by the teacher modelling the equation writing until the student is ready.
Subitizing	**Subitizing** Teach the dice patterns 4-6. Explore the patterns they can see within e.g. 4 is 2 and 2, 6 is 3 and 3, 5 is 4 with one in the middle or diagonal lines of 3. Then roll a 1-6 dice and recognise the pattern quickly without counting. For further practice: Fingers – how many fingers am I holding up on 1 hand, can you hold up X fingers? Can you do it in a different way Subitize picture cards of hand R14
Counting	**Counting – patterns and objects** R2 or R3 LF M4 V2 Count to 10 – put out number rainbow and matching pictures, include number words - colour coded, turn over number cards R4 & R7

Counting continued	Trace and write numbers to 0-6 in multiple ways (trace, draw, air write etc) and then copy onto a number track R5 Use hand model
Number Sense	**Number sense focus area explain and model Examples and nonexamples** LF M4 V3 Practise 1:1 counting of groups of items to 6. Explore ways to check the count – recount, count in a different way, move and count 1 at a time. LF M4 V4 Compare 2 groups of items to 6 and explore the concepts of more/less/same. How do you know? How can you prove it?
Games	**Game/ hands on activity** LF M4 V5 War numbers 0-6 Use the cards 0-6 from R2 or R3 . Shuffle and place the cards in the centre. Each player takes 1 card. The player with the highest number wins. You can change the game up with smallest number or closest to another number e.g. closest to 3.
Word Problems	**Word Problems – Problem Solving Book Foundation Level** – Problem solving tasks will need to be supported with concrete and representative materials – e.g. counters and drawing pictures. Read the question to the student, possibly more than once. What does the question ask you to do? How can you solve this?

 Number Problems	**A: Number problems** – solve the problem in different ways. These start from Module 5.
 Games	**Game/ hands on activity** [R11] Revisit the Lady Beetle spots game now using numbers 1-6 using a 6-sided dice. Play solo or against a partner. LF M4 V6  Addition: Roll the dice, draw the dots on one side of the ladybug. Roll again to complete the second side. Add the total together. On a second ladybug sheet, the second player can do the same. The winner is the player with the most or least – players to decide the rules. LF M4 V7 Subtract the number rolled on the dice. Use a 1-6 dice and roll the dice. The winner for each round is the player who has the least number of dots left.

Module 5 – Addition – concrete to recording.
Ideas for teacher lesson (1 hour)

Thinking Problems	**Thinking task** R1 LF M5 V1 Keep or change Roll a 0-3 dice. Write the first number you roll. Then roll again – you can keep or change (reroll) the dice – the aim being to get the highest number from the 2 rolls. For example, if roll 1 was 3 and roll 2 as a 0, you may wish to roll the second dice again and try to get a higher number. However, if the second number was a 2 would you roll again?? As they play, record the addition equation in standard form. You can re play the game with a 6-sided dice and work on subtraction problems here. Roll both dice together and place for subtraction with the highest number first. Allow 1 reroll of one of the dice to improve the score. Here you are competing for the lowest answer. Again, record the number equation in standard form.
Subitizing	**Subitizing** LF M5 V2 Dice 1-6 patterns review. Write the numbers 1-6 on a page. Which player is first to roll all 6 numbers?
Counting	**Counting – patterns and objects** R2 or R3 LF M5 V3 <u>Count to and from 10</u> – put out number rainbow and matching pictures, include number words - colour coded if still needed – if not just use plain cards, practice counting forwards and backwards 0- 10

Number Sense	**Number sense focus area explain and model Examples and nonexamples** Teach adding 1 more. Throw a ball or similar. One player says a number, the other says 1 more. You can also extend to 2 more. R9 **Teach symbols +-= and their actions** R15 LF M5 V4 Teach joining sums orally and show how to record in writing. Stay within numbers to 6. Part part whole and joining sums
Games	**Game/ hands on activity** R16 LF M5 V5 Go fish or memory addition game. Find the answer to your question. Numbers 1-6 only. There are multiple questions with the same answer but there are enough cards to match them all up.
Word Problems	**Word Problems – Problem Solving Book Foundation Level** – Problem solving tasks will need to be supported with concrete and representative materials – e.g. counters and drawing pictures. Read the question to the student, possibly more than once. What does the question ask you to do? How can you solve this? Track which types of questions the students can and can't do.

 Number Problems	A: **Number problems** – solve the problem in different ways. These start from Module 5. Write out one question at a time into a separate book (the question booklet is for teacher reference) **Conceptual Understanding Pentagon** 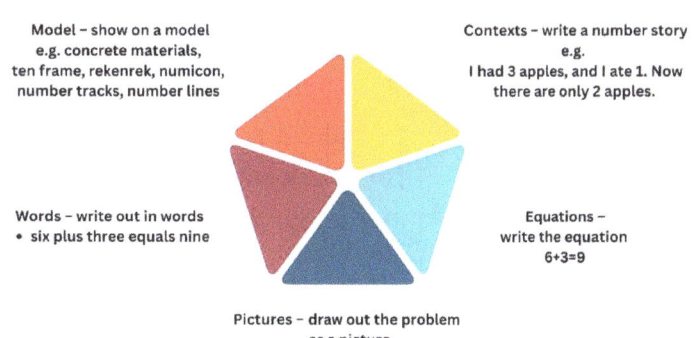 Model – show on a model e.g. concrete materials, ten frame, rekenrek, numicon, number tracks, number lines Contexts – write a number story e.g. I had 3 apples, and I ate 1. Now there are only 2 apples. Words – write out in words • six plus three equals nine Equations – write the equation 6+3=9 Pictures – draw out the problem as a picture Van de Walle 2006, adapted Mike Flynn B: **Retrieval and interleaving practice tasks –** Foundation Level book starting at Module 7.
 Games	Game/ hands on activity R1 & R17 LF M5 V6  Forwards backwards game Roll the dice and move your counter. If you land on a question move to the answer. This works best with a 1-3 dice.

Module 6 – Subtraction – concrete to recording.

Ideas for teacher lesson (1 hour)

Thinking Problems	**Thinking task** LF M6 V1 Keep or change Let's replay the keep or change game from Module 5 but with a 6-sided dice and work on subtraction problems here. Roll 2 dice together and place for subtraction with the highest number first. Allow 1 reroll of one of the dice to improve the score, this may mean you need to change the order of the dice to make a solvable equation. Here you are competing for the lowest answer. Record the number equation in standard form.
Subitizing	**Subitizing** R8 5 frames - use all frames 0-5 Take time to look at each card and count up the turtles. Once the student is familiar with these explore which they can subitize. How fast can they do them all? Can they beat their time for how many they can recognise without consciously counting?
Counting	**Counting – patterns and objects** R14 Write numbers to 1-6 in any order from finger subitizing cards R4 use tracing sheet if needed– find the number and trace. R7 Count back from 6 using number track

Counting continued	<u>Trace and write</u> numbers to 0-6 in multiple ways e.g. onto whiteboard, air write, paint with water, chalk etc. R5 Use hand model
Number Sense	**Number sense focus area explain and model Examples and nonexamples** Teach 1 less Take away up to 6 from the group – include answers of zero Throw a ball or similar. One player says a number, the other says 1 less. You can also do 2 less as the student progresses. R18 LF M6 V2 Teach subtraction sums orally and show how to record in writing. Stay within numbers to 6.
Games	**Game/ hands on activity** R19 LF M6 V3 Forwards backwards game Roll the dice and move your counter. If you land on a question move to the answer. This works best with a 1-3 dice R1
Word Problems	**Word Problems – Problem Solving Book Foundation Level** – Problem solving tasks will need to be supported with concrete and representative materials – e.g. counters and drawing pictures. Read the question to the student, possibly more than once. What does the question ask you to do? How can you solve this?

 Number Problems	A: **Number problems** – solve the problem in different ways. These start from Module 5. Write out one question at a time into a separate book (the question booklet is for teacher reference) **Conceptual Understanding Pentagon** 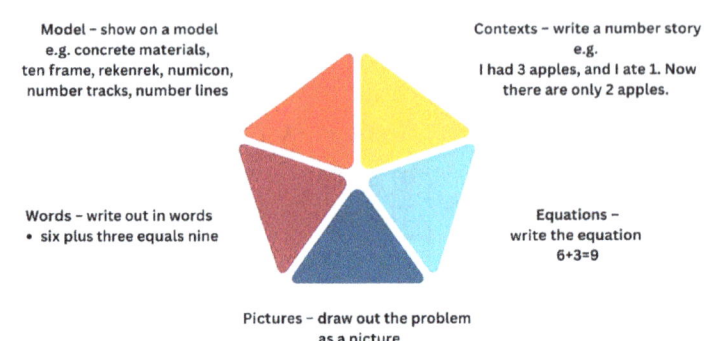 Model – show on a model e.g. concrete materials, ten frame, rekenrek, numicon, number tracks, number lines Contexts – write a number story e.g. I had 3 apples, and I ate 1. Now there are only 2 apples. Words – write out in words • six plus three equals nine Equations – write the equation 6+3=9 Pictures – draw out the problem as a picture Van de Walle 2006, adapted Mike Flynn B: **Retrieval and interleaving practice tasks** – Foundation Level book starting at Module 7.
 Games	**Game/ hands on activity** <u>R20</u>  LF M6 V4 Go fish or memory subtraction game. Find the answer to your question. Numbers 1-6 only. There are multiple questions with the same answer but there are enough cards to match them all up.

Module 7 – Doubling concept.
Ideas for teacher lesson (1 hour)

Thinking Problems	**Thinking task** R2 or R3 LF M7 V1 Draw 2 number cards out of a pile. Which number is bigger (or smaller)? How do you know? Can you prove it? Use concrete materials or pictures to show. One way would be to line up the numbers side by side or stack them up. 6 is more than 3. Or 3 is less than 6 5 is more than 3. Or 3 is less than 5.

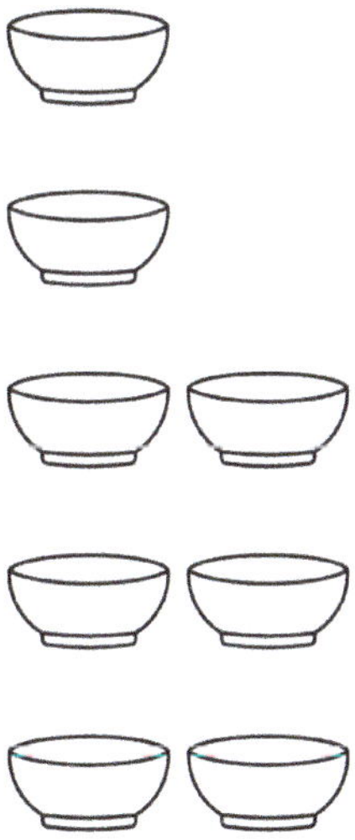

	**Subitizing** Subitize with Dominoes to 6. Look for doubles patterns too. Doubles Not doubles  
	**Counting – patterns and objects** LF M7 V2  Teach 1st, 2nd, 3rd, middle, last. Line up a row of animals or similar 5 or more works well. Explore which is 1st, 2nd, 3rd, middle, last. Move them around and repeat. You could even add a few more or take some out.  R4 & R5 Teach students to write numbers 7,8,9,10. R14 Practice writing numbers 0 to 10 in any order from finger subitizing cards, use tracing sheet if needed – find the number and trace.

 Number Sense	**Number sense focus area explain and model Examples and nonexamples** R21 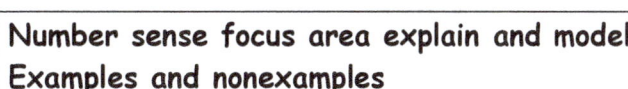 LF M7 V3 Doubles bus 1-5 version - place the same number on the top of the bus as the bottom of the bus to demonstrate doubles  Noah's Ark – the animals came in pairs = doubles Make your own ark picture with doubles of each animal 
 Games	**Game/ hands on activity** R22  LF M7 V4 Forwards backwards game doubles Roll the dice and move your counter. If you land on a question move to the answer. This works best with a 1-3 dice R1

 Word Problems	**Word Problems – Problem Solving Book Foundation Level** – These start at Module 2. Problem solving tasks will need to be supported with concrete and representative materials – e.g. counters and drawing pictures. Read the question to the student, possibly more than once. What does the question ask you to do? How can you solve this?
 Number Problems	**A: Number problems** – solve the problem in different ways. These start from Module 5. Write out one question at a time into a separate book (the question booklet is for teacher reference) **Conceptual Understanding Pentagon** 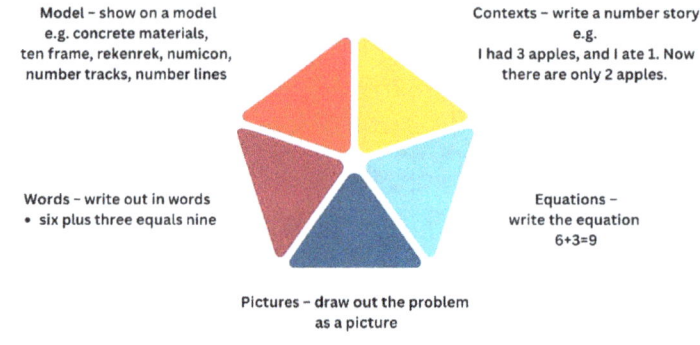 Van de Walle 2006, adapted Mike Flynn **B: Retrieval and interleaving practice tasks – Foundation Level book starting at Module 7.**

	**Game/ hands on activity** R1 & R23 LF M7 V5  Skittles game – roll a dice numbers 0-3 and knockout the skittle which is double the number. Write your chosen numbers on the skittles 0-3. Cross them off once you have rolled the number and successfully doubled it.

Module 8 – Sharing concept.
Ideas for teacher lesson (1 hour)

Thinking Problems	**Thinking task** Making Pizza LF M8 V1 Ask students roll a 1-6 dice twice. The first roll tells them how many pizzas to draw. The second roll tells them how many pepperonis to put on EACH pizza. Then they write the number sentence that will help them answer the question, "How many pepperonis in all?"
Subitizing	**Subitizing** R14 Fingers to 10. How fast do you know how many fingers? Encourage 5 plus some more thinking for 2 hands.
Counting	**Counting – patterns and objects** R4, R5, R14 Write numbers to 10 in any order from finger subitizing cards, use tracing sheet and hand formation prompt card if needed – find the number and trace. Use the full hand card plus other digits to show 6-10.
Number Sense	**Number sense focus area explain and model Examples and nonexamples** LF M8 V2 Share items equally between 2 (or 3) groups/ 2 (or 3) people – 1 for me, 1 for you strategy (maximum 12 items). Count to check groups are equal

© Copyright 2025 Mathtastic: Tracy Ashbridge. All rights reserved

Number sense continued	Share out until no more left, if some left over discuss what could be done in everyday situations.
Games	**Game/ hands on activity** R24 LF M8 V3 Teddy bears picnic game Ask students roll a 1-6 dice twice. The first roll tells them how many Teddies are in play. The second roll tells them how many cakes to put/draw on each place. After all players have had their go the winner is the person who has the most on a plate with the least left over.
Word Problems	**Word Problems – Problem Solving Book Foundation Level** – Problem solving tasks will need to be supported with concrete and representative materials – e.g. counters and drawing pictures. Read the question to the student, possibly more than once. What does the question ask you to do? How can you solve this?
Number Problems	**A: Number problems** – solve the problem in different ways. These start from Module 5. Write out one question at a time into a separate book (the question booklet is for teacher reference)

Number problems conttinued	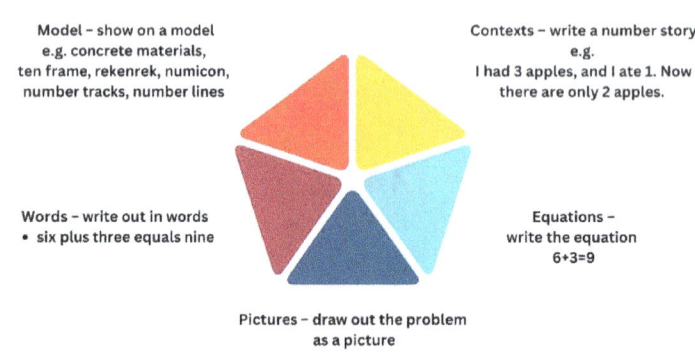

Van de Walle 2006, adapted Mike Flynn

B: Retrieval and interleaving practice tasks – Foundation Level book starting at Module 7. |
|  | **Game/ hands on activity**

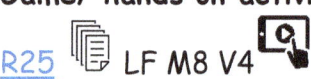

Sharing game. Roll the dice and move around the track. Solve the problems. The adult will need to read the questions. Have counters or similar available to practically share out the items for the questions. For questions where there is not an equal share discuss possible options. |

Foundation Numbers to 1-6 Resources

Resources – not included
- Dice 1-6
- Blank 6-sided dice
- Dominoes
- Counters

Lesson Plan

	Thinking task
Thinking Problems	
Subitizing	Subitizing
Counting	Counting – patterns and objects
Number Sense	Number sense focus area explain and model Examples and nonexamples
Games	Game/ hands on activity

	**Word Problems – Problem Solving Book Foundation Level** – these start at Module 2. Problem solving tasks will need to be supported with concrete and representative materials – e.g. counters and drawing pictures. Read the question to the student, possibly more than once. What does the question ask you to do? How can you solve this?
	**A: Number problems** – solve the problem in different ways. These start from Module 5. Write out one question at a time into a separate book (the question booklet is for teacher reference) **Conceptual Understanding Pentagon** 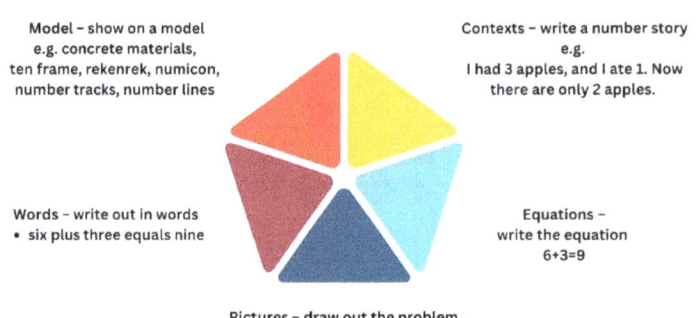 Van de Walle 2006, adapted Mike Flynn **B: Retrieval and interleaving practice tasks – Foundation Level book starting at Module 7.**
	Game/ hands on activity

Homework - 15 mins per day

Day 1	
Counting	
Number Problems	
Games	
Day 2	
Counting	
Problem Solving	Word Problem solving booklet – Foundation Level
Games	

Day 3	
Counting	
Number Problems	Number problems booklet Foundation Level
Games	
Day 4	
Counting	
Problem Solving	Word Problem solving booklet – Foundation Level
Games	

📑 Resource 1 Net for cube to make 1-3 dice (or use blank sided dice available from craft sections of discount stores)

Net for cube to make 1-3 dice

Resource 2 Number cards – coloured

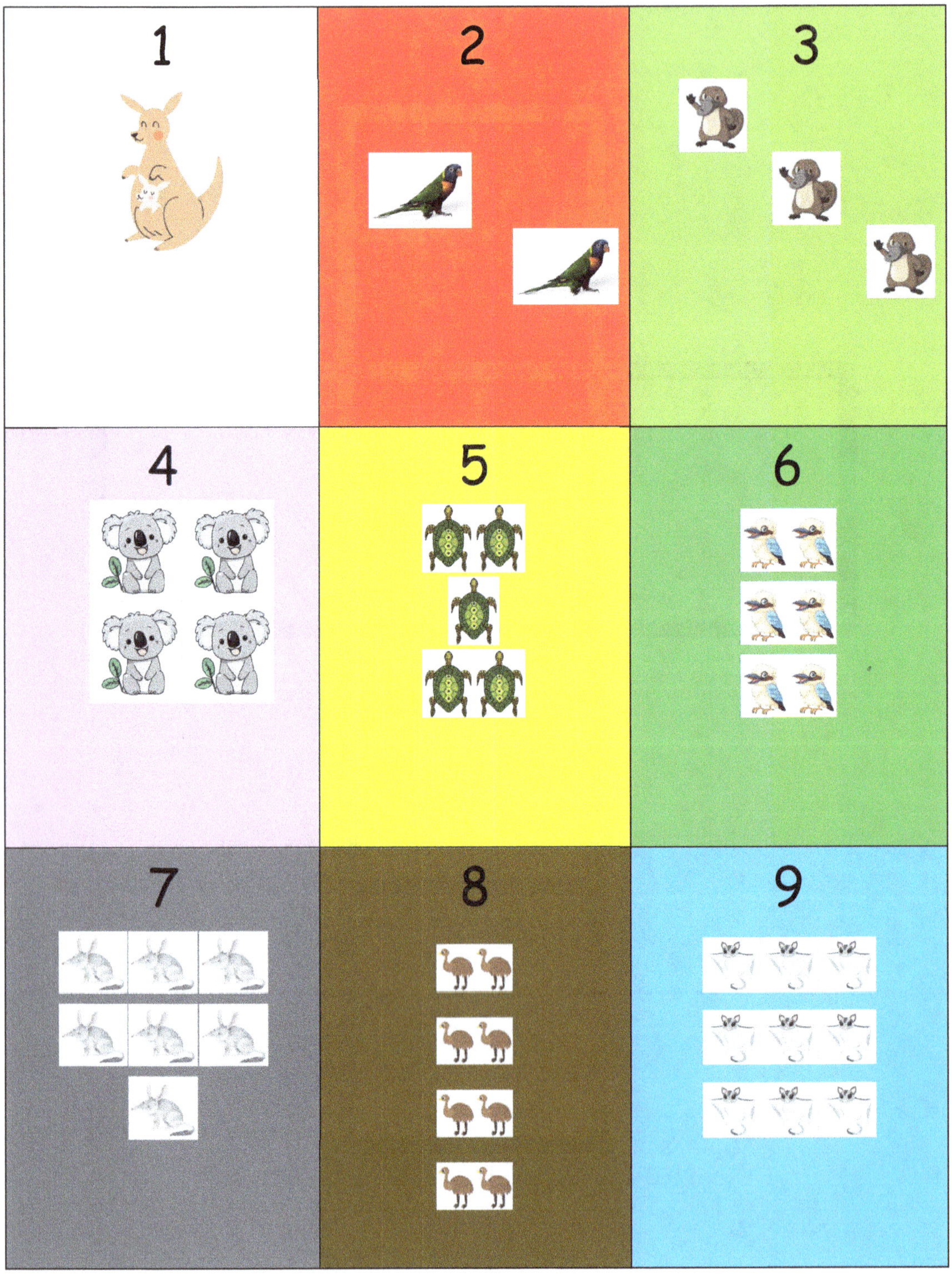

10  🦔🦔 🦔🦔 🦔🦔 🦔🦔 🦔🦔	one	two
three	four	five
six	seven	eight

nine	ten	1
2	3	4
5	6	7

0 zero

Resource 3 Number cards – no colour

1 	2 	3 
4 	5 	6 
7 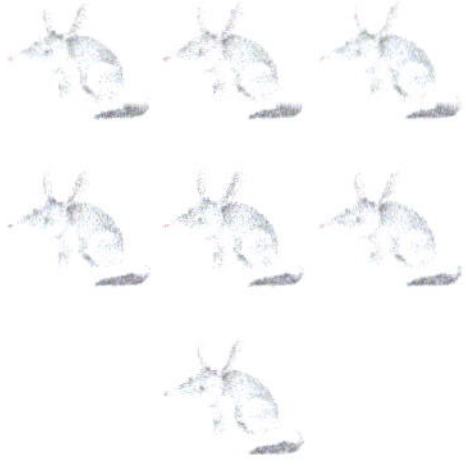	8 	9 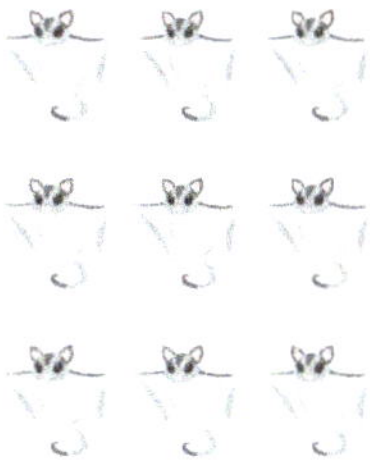

10 (ten hedgehogs)	one	two
three	four	five
six	seven	eight

nine	ten	1
2	3	4
5	6	7

8	9	10

(7 bilbies)	(10 emus)	(9 possums)
(12 echidnas)	0	zero

 Resource 4 Tracing Numbers

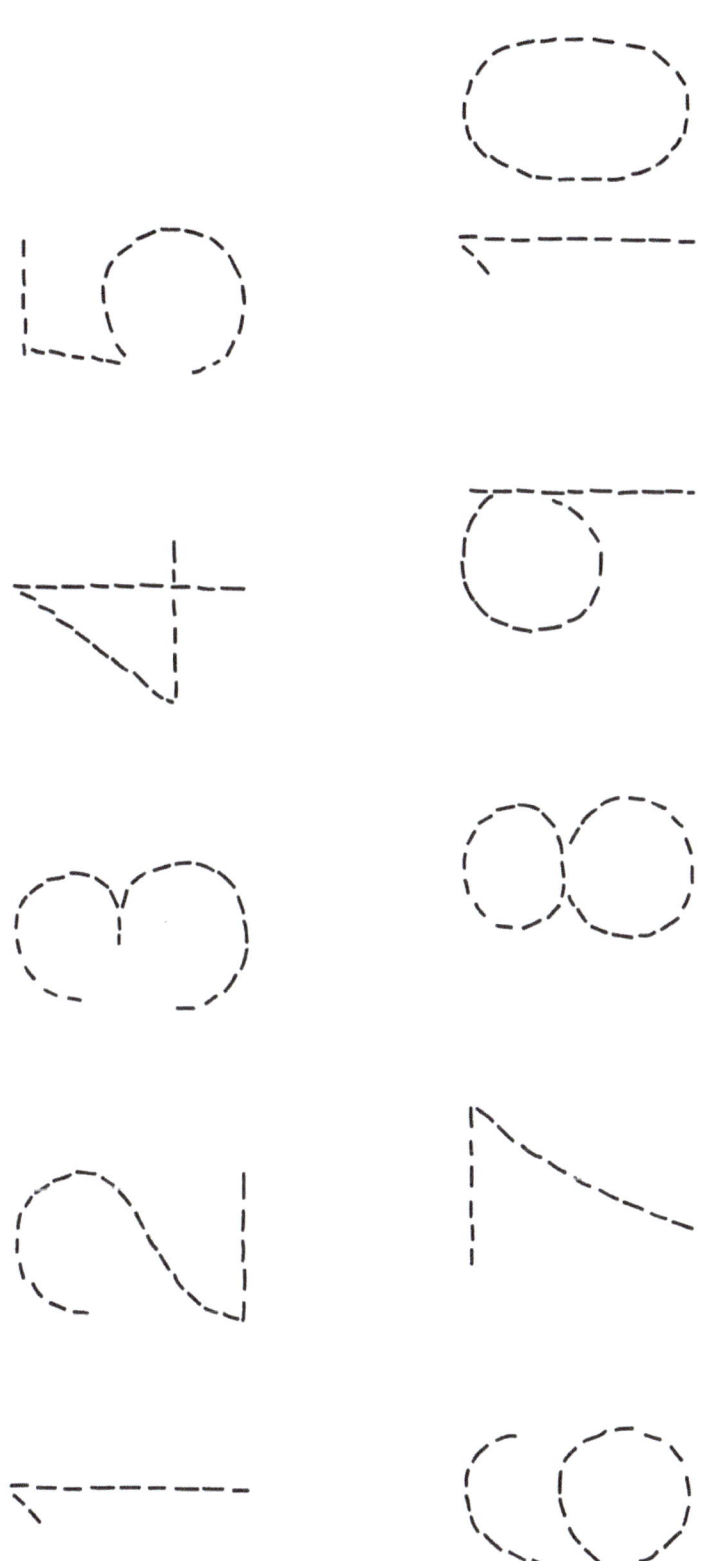

Resource 5 Number formation reference chart using the hand

How to write numbers

Resource 6 Memory Game 0-3 dice patterns

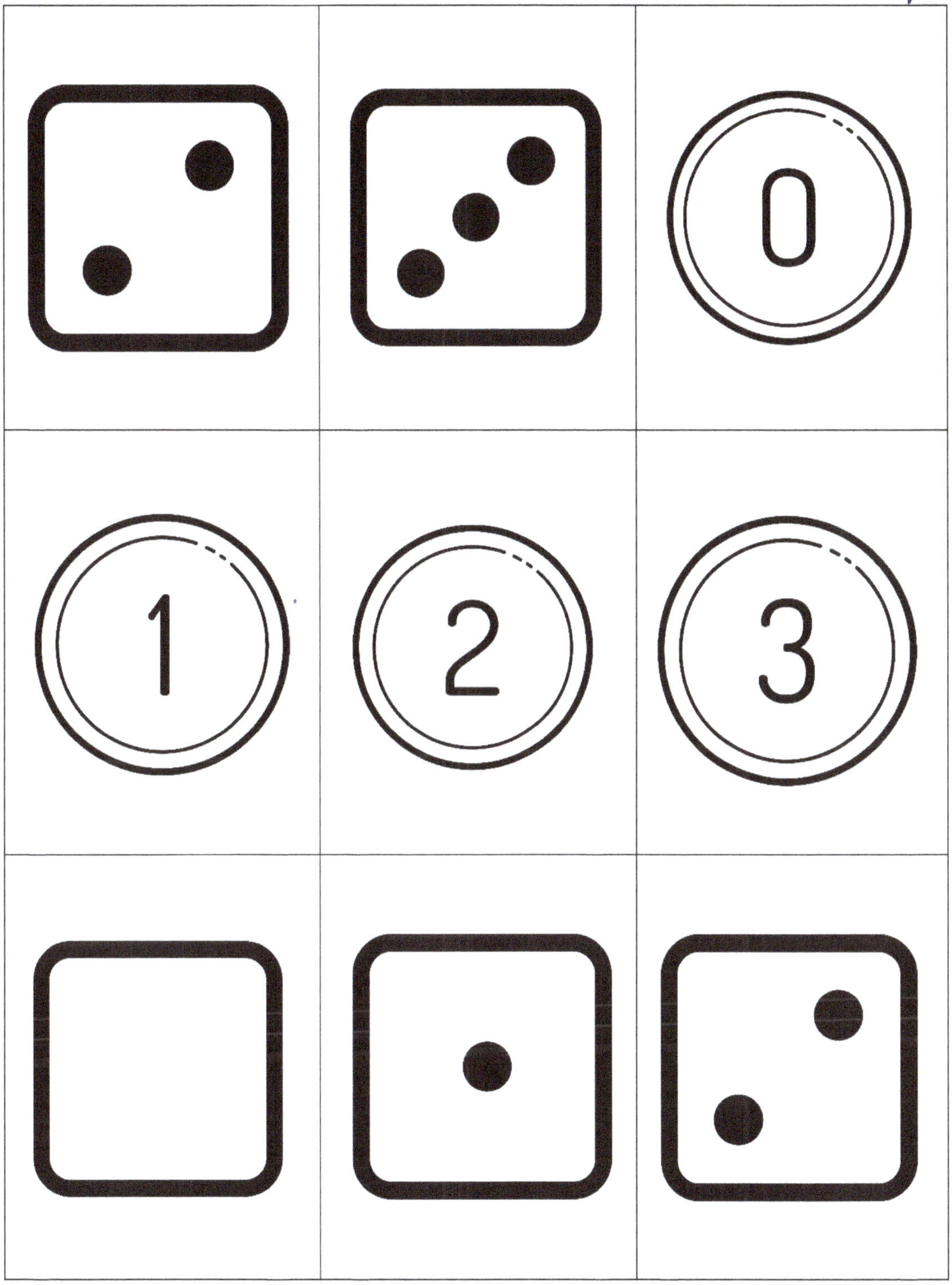

Resource 7 Number Track Game

0	1	
1	2	
2	3	
3	4	
4	5	
5	6	
6	7	
	8	
	9	
	10	

Resource 8 Five Frames – 5 wise and random

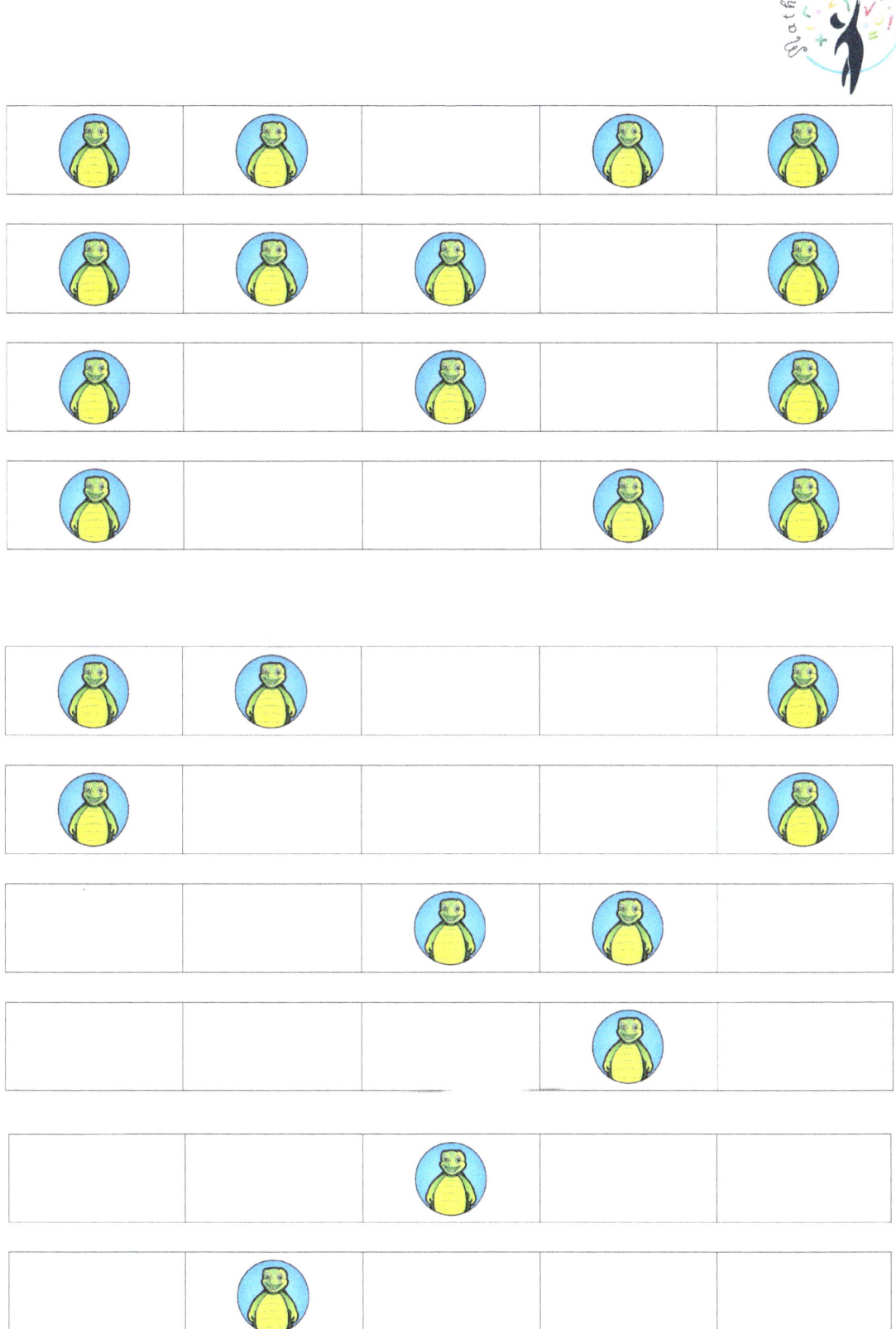

Resource 9 Addition and Subtraction actions

Addition
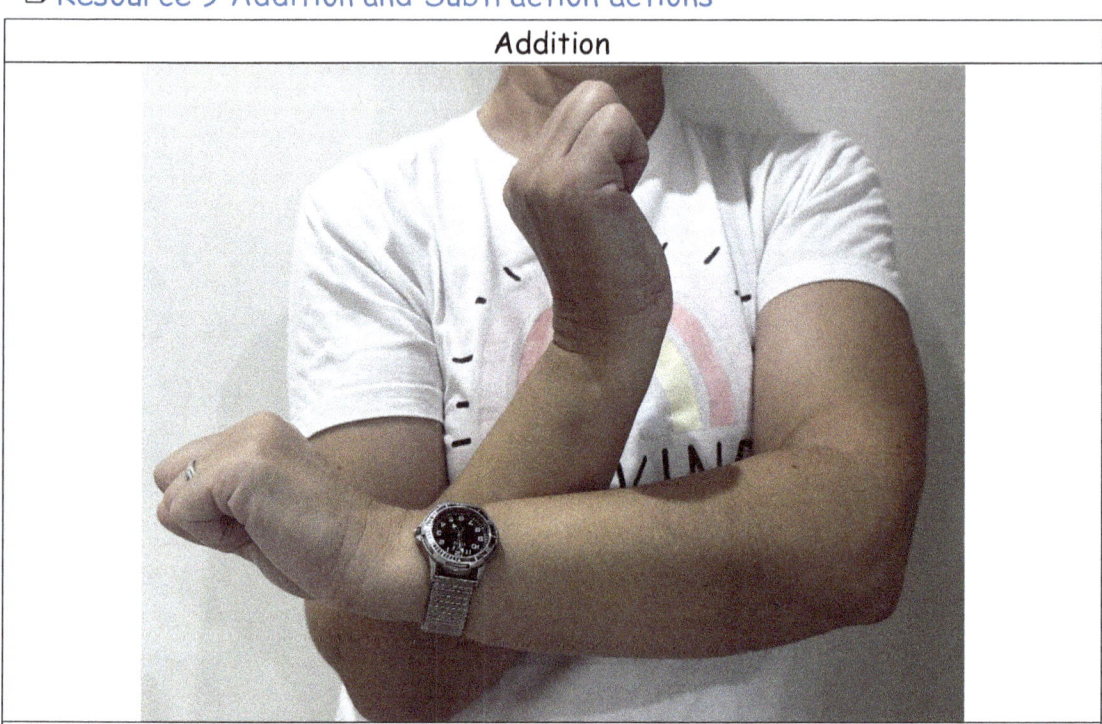
Subtraction

Resource 10 Cover Up Game

COVER UP GAME
1	2	3
0	3	5
6	2	4

COVER UP GAME
0	1	3
4	1	5
6	2	4

COVER UP GAME
6	5	3
0	4	5
6	2	4

COVER UP GAME
3	1	3
0	6	5
6	2	4

Resource 11 Lady Beetle Spots Game (laminate or place into a plastic sleeve)

Resource 12 Cover up number formation game cards – cut into 4 playing cards (laminate or place into a plastic sleeve)

0	1	2	3	1	2	3	0
0	2	1	0	0	1	0	1
0	3	2	3	1	3	1	2
2	1	0	3	0	2	2	1
1	3	2	0	2	0	2	3
2	2	1	1	3	1	0	0
3	2	3	2	1	1	3	1
3	0	0	3	2	3	0	2

Resource 13 Snakes and Ladders

SNAKES & LADDERS

GOAL:

To be the first player to make their way to 100.

HOW TO PLAY:

Place counters at the start. Players roll a die numbered with 0,1,2,3 only (Resource 1) and move that number of spaces on the board. If a player lands on a ladder, they can climb to the top of it. If they land on a snake, they need to slide back down to the bottom.

Resource 14 Finger subitizing

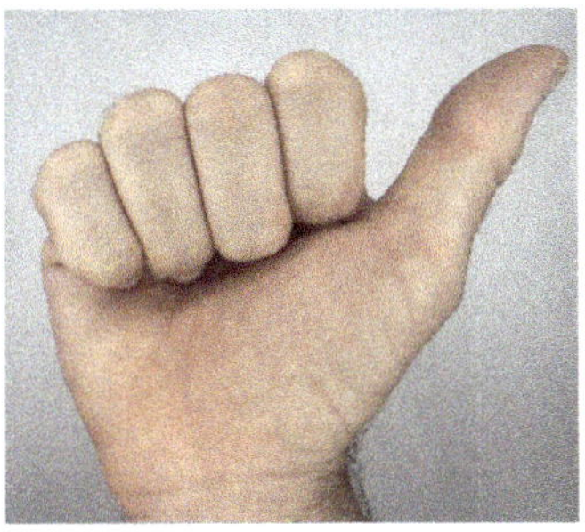

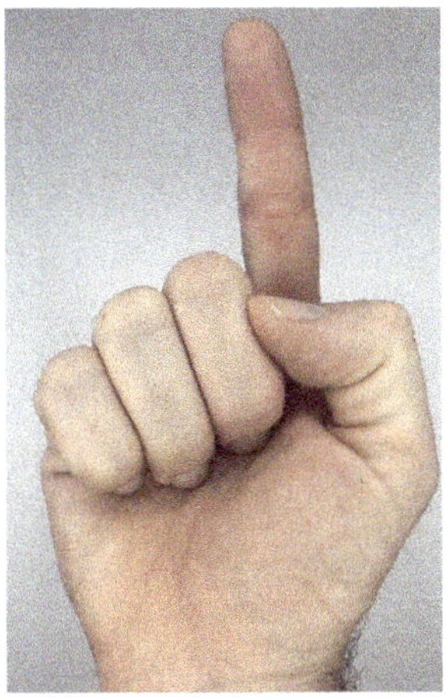

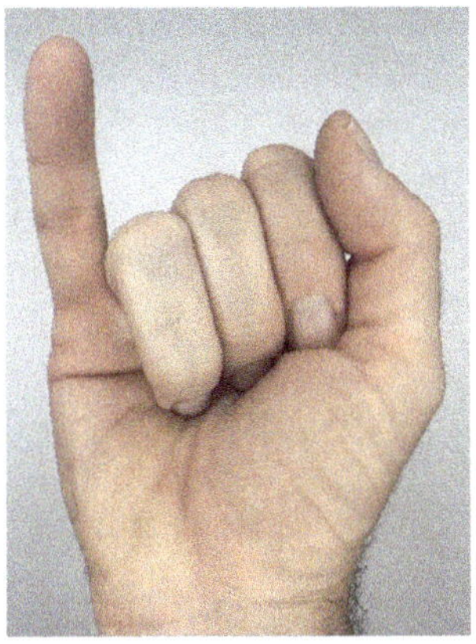

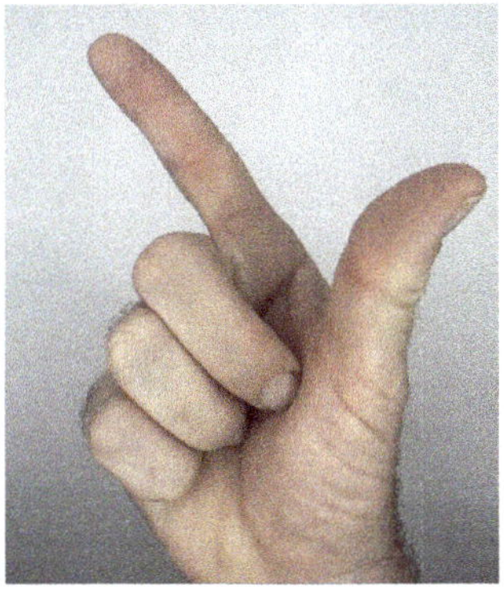

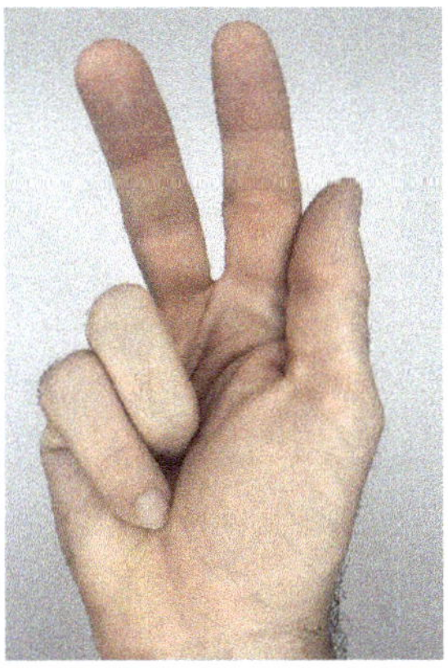

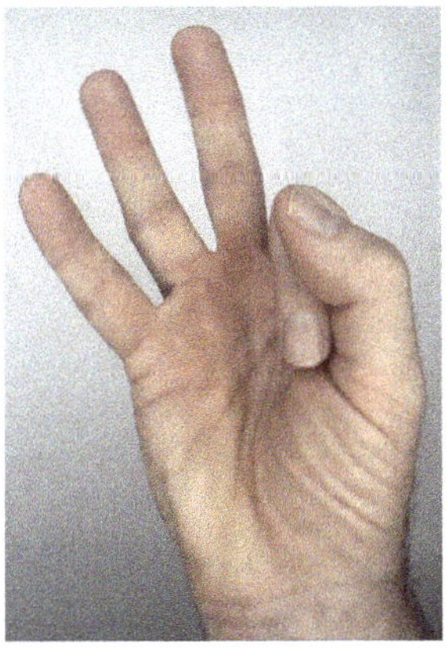

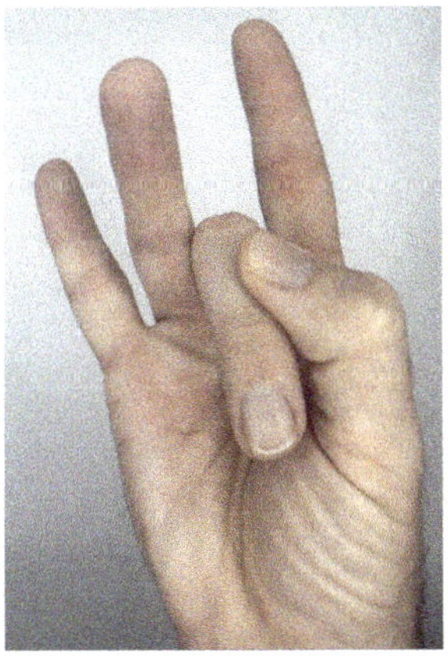

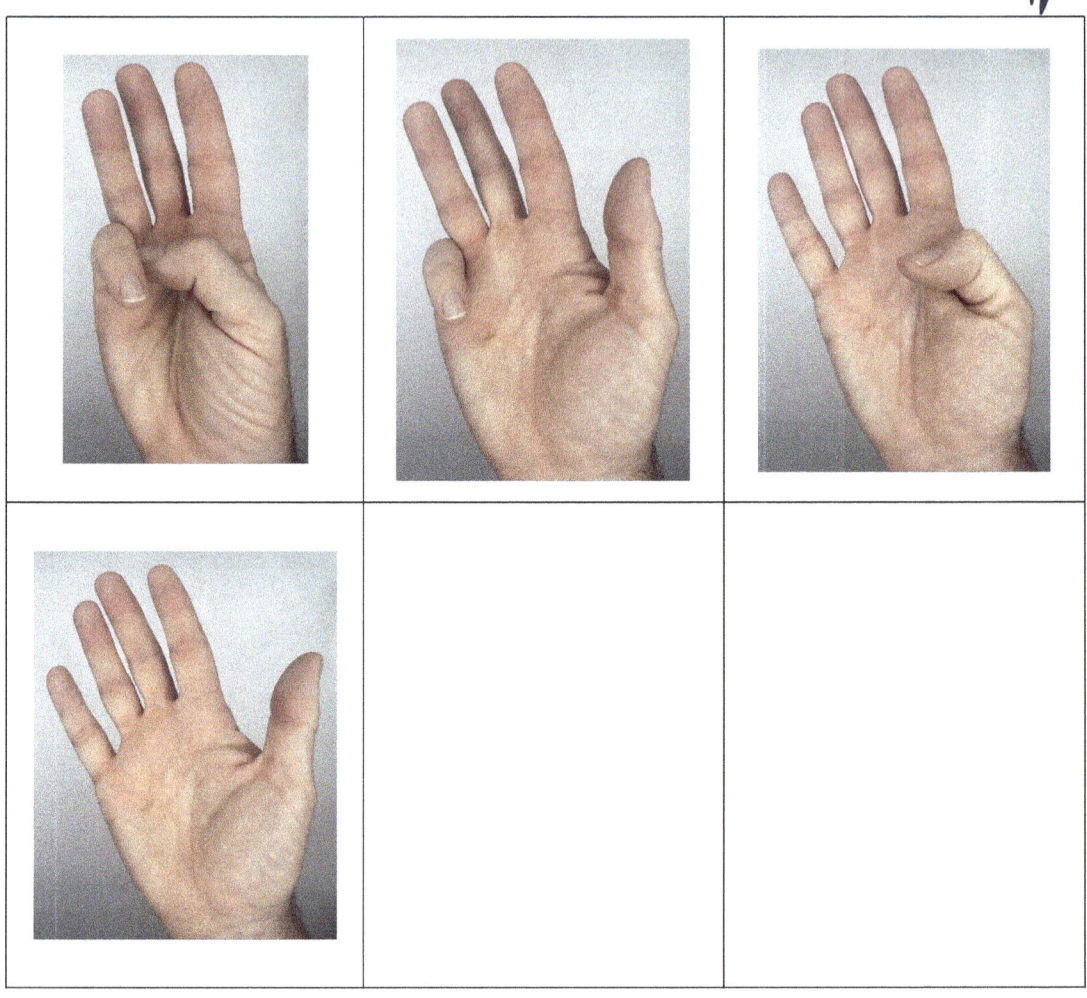

Resource 15 Addition prompt page

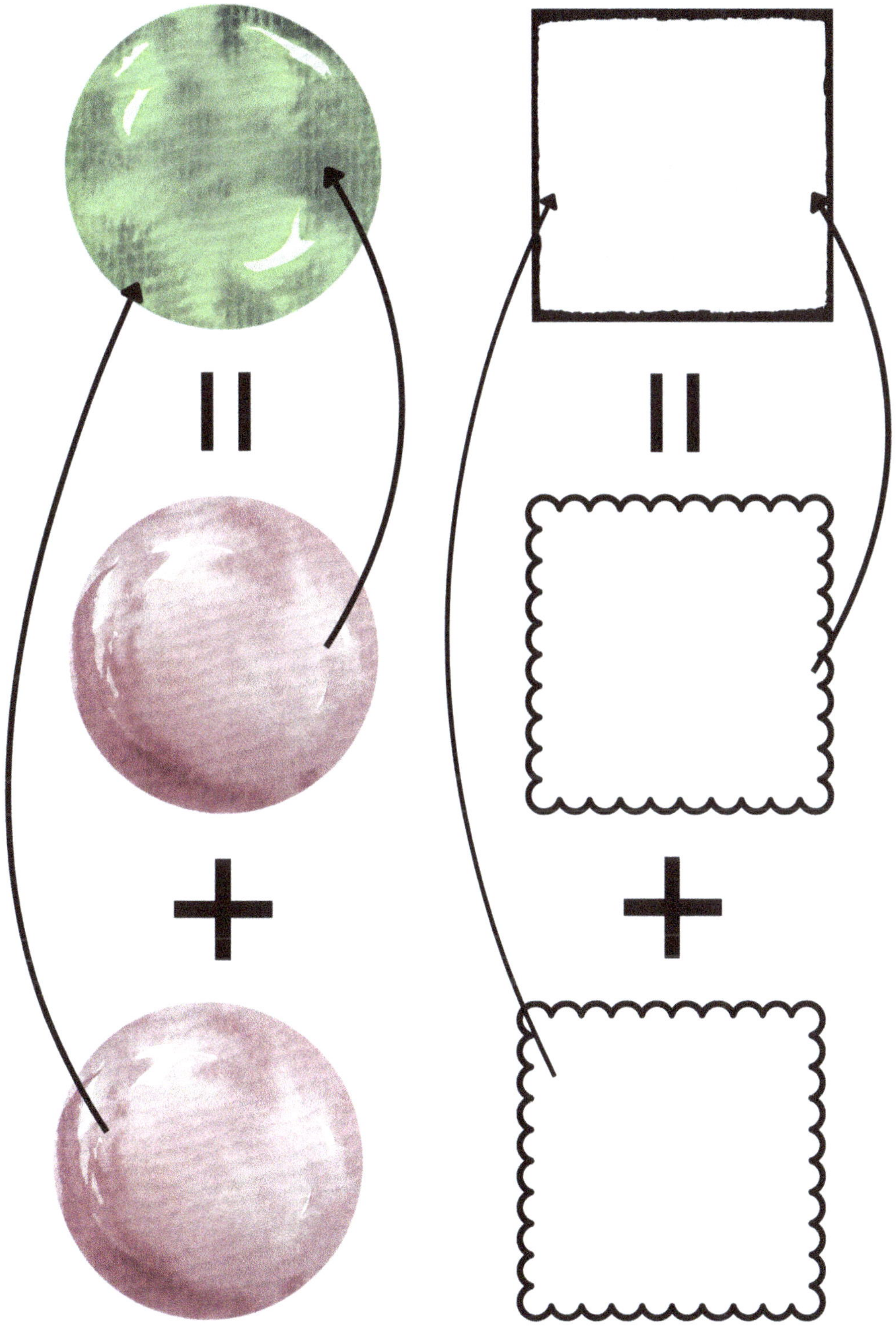

Resource 16 Go Fish addition to 6

6+0	5+1	4+2
3+3	6	6
6	6	5+0

4+1	3+2	5
5	5	4+0
3+1	2+2	4

4	4	3+0
2+1	3	3
2+0	1+1	2

| 2 | 1+0 | 1 |

Resource 17 Addition Forwards Backwards Game

FORWARDS BACKWARDS GAME

Roll the dice and move your counter. If you land on a question, move to the answer.

Board spaces (clockwise from START):
START → 1+1 → 3 → 2 → 2+2 → 3+3 → 6 → Roll again → 4 → 2+1 → 0 → 1+0 → Miss a turn → 0+0 → 1 → 5 → 4+1 → 3+1 → FINISH

Resource 18 Subtraction Prompt Page

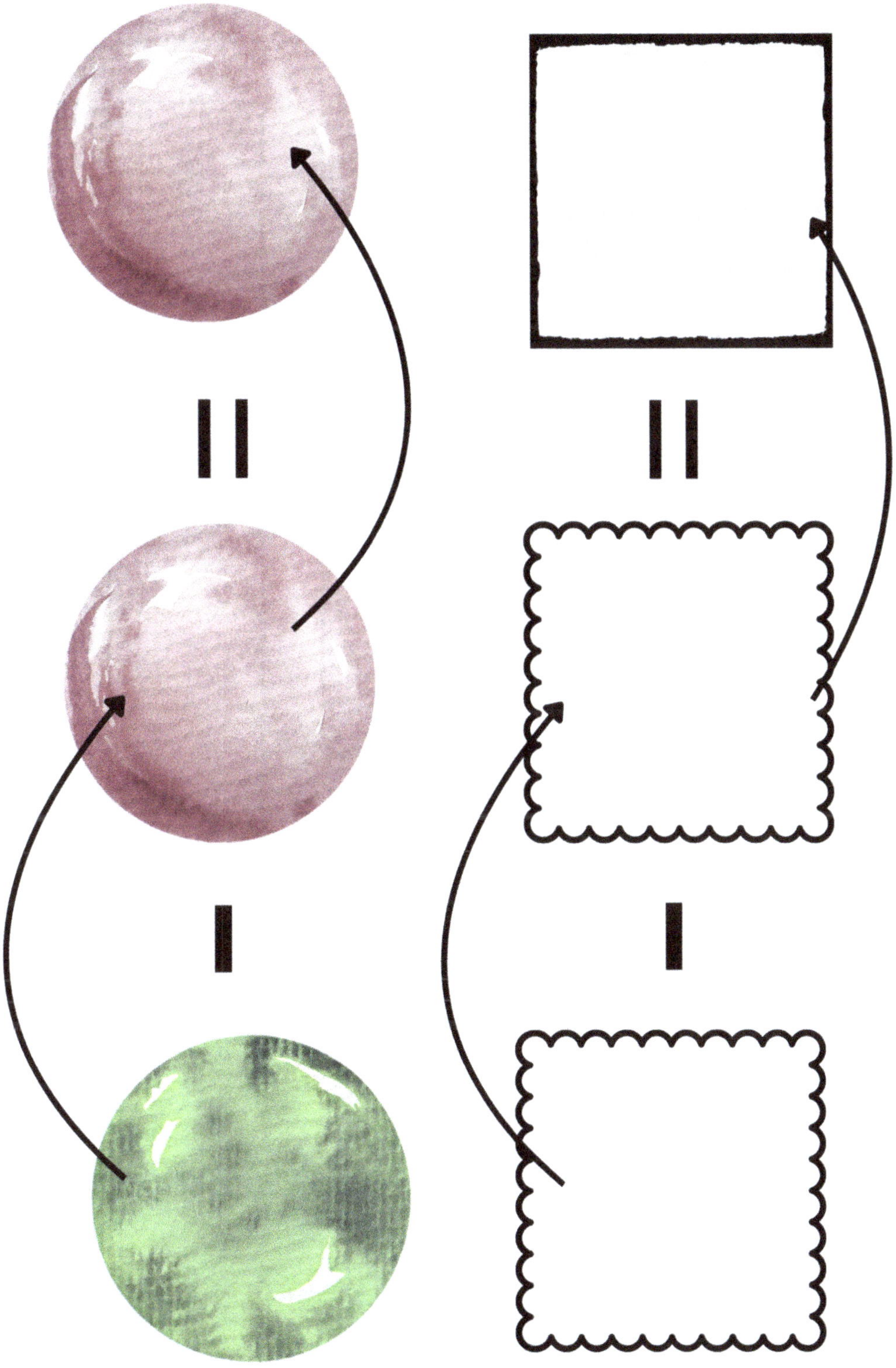

Resource 19 Subtraction Forwards Backwards Game

FORWARDS BACKWARDS GAME

ROLL THE DICE AND MOVVE YOUR COUNTER. IF YOU LANND ON A QUESTTION, MOVE TO HE ANSWER

| START | 6-1 | 4-2 | 5 | 3-3 | 5-4 | 3 |

FINISH						Roll again
6						2
5-5	4	1	6-2	Miss a turn	0	4-1
						5-2

Resource 20 Subtraction Go Fish or Memory Game

6-6	6-5	6-4
6-3	6-2	6-1
0	1	2

3	4	5
6-0	6	5-5
5-4	5-3	5-2

5-1	5-0	0
1	2	3
4	5	4-4

4-3	4-2	4-1
4-0	0	1
2	3	4

3-3	3-2	3-1
3-0	0	1
2	3	2-2

2-1	2-0	0
1	2	1-1
1-0	0	1

Resource 21 Doubles Bus

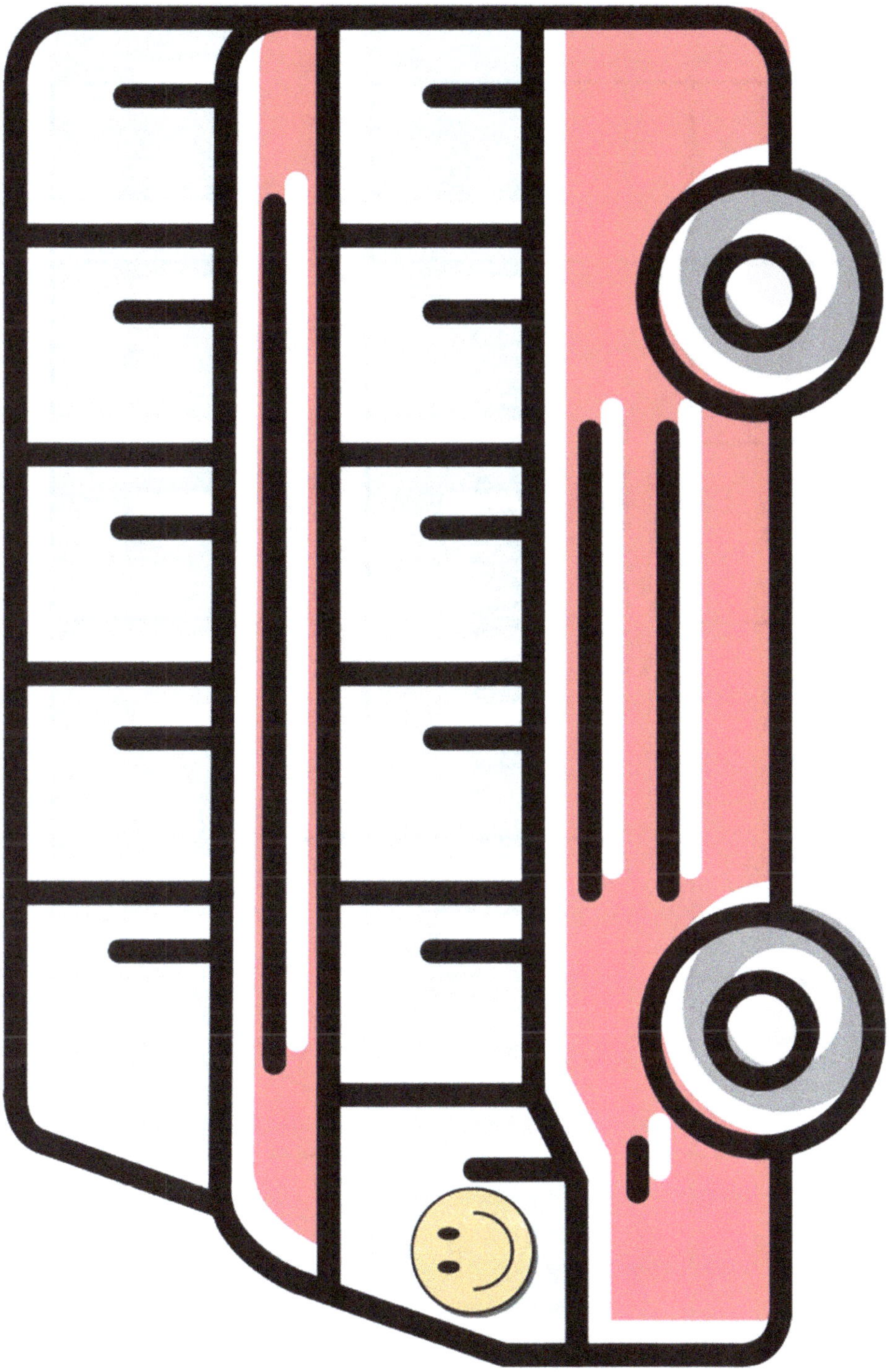

Resource 22 Forwards backwards game doubles

FORWARDS BACKWARDS GAME

DOUBLES

Roll the dice and movve your counter. If you lannd on a questtion, move to he answer

Board spaces
START
0+0
3+3
0
3+3
2+2
6
Roll again
2
1+1
2+2
3+3
Miss a turn
1+1
4
Rest here
FINISH
1+1
2+2

Resource 23 Skittles game (laminate or place in plastic sleeve)

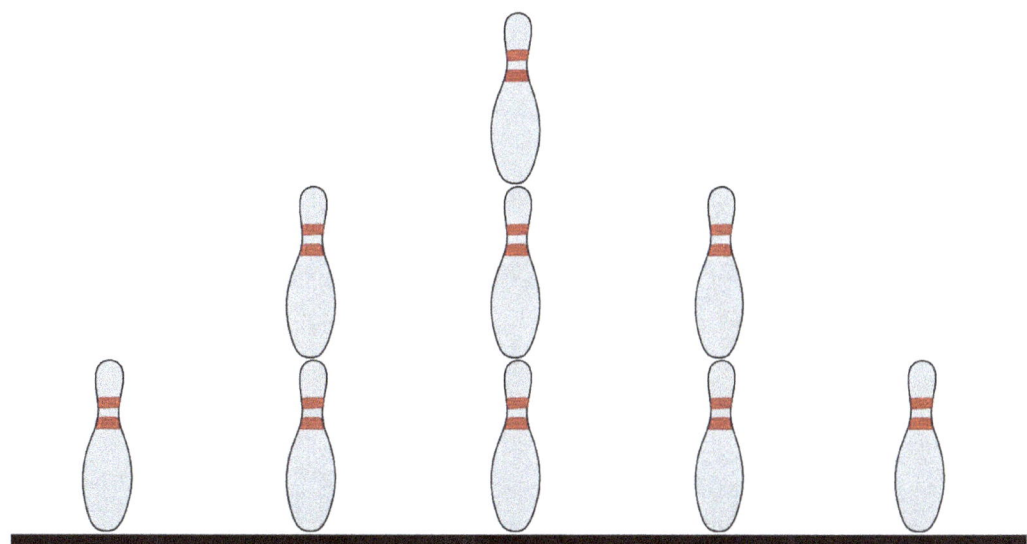

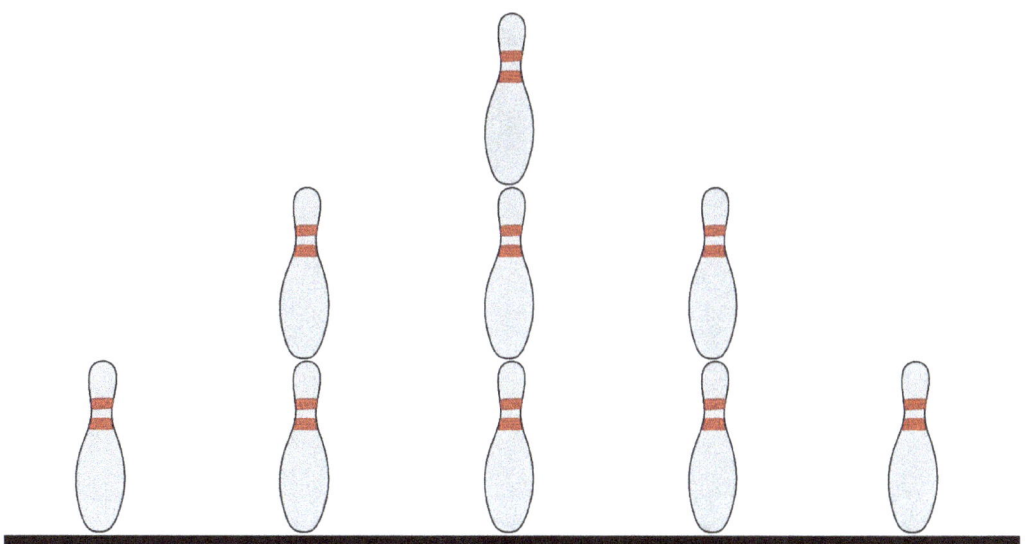

📑 **Resource 24 Teddy Bears Picnic** (Print on A3, laminate or place in plastic sleeve)

Teddy Bears Picnic Game

Resource 25 Sharing Game

SHARING PROBLEMS

Roll the dice and move the correct number of spaces.
Solve the problem you land on. Use concrete materials to help.

START

FINISH

Board spaces (in play order from START):
- Share 4 apples between 2 children
- Share 5 bones between 2 dogs
- Share 3 coins between 3 sisters
- Share 2 oranges between 2 friends
- Share 6 paint brushes between 2 children
- Share 5 crayons between 5 girls
- Share 3 books between 2 boys
- Go back 3 spaces!
- Share 4 lollipops between 3 children
- Share 6 grapes between 3 children
- Sing the chorus of your favorite song.
- Miss a turn!
- Share 3 bananas between 2 children
- Share 5 cheese sticks between 2 girls
- Share 5 fish between 2 bowls
- Bonus turn! Roll the dice again.
- Share 1 carrot between 2 rabbits
- Share 6 Lego pieces between 2 children
- Share 4 chocolates between 4 teachers
- Share 3 marshmallows between 3 people
- Share 6 frogs between 2 ponds

© Copyright 2025 Mathtastic: Tracy Ashbridge. All rights reserved

Foundation Numbers to 1-6 Workbook

Instructions

This booklet can be printed from the resource page on the website – see Page 2.

For students in the early stages of learning maths it is important to talk about maths orally and carry out maths supported by concrete and pictorial means before abstract representations. For this reason, solving number problems presented in standard format (e.g. 4+2=) are only introduced from Module 5. For earlier modules, use the worded problems to practice the addition and subtraction concepts.

From Module 5 answer the questions on each page. Each row is for a different day. The first column contains practice related to the work in the module and the second column is for practice from previous modules. Problems should be presented alternatively horizontally and vertically from Module 2 onwards.

Students (or adults) should copy the problem into a maths exercise book – this can be plain, lined or squared. In the early stages a plain book may be easier.

Divide the page into 2 columns – the first for writing the problem and the second to represent in the 5 different ways. See below for an example.

For each of the questions in the first column solve the problem and note how it was solved in the third column – see below. Then show understanding using one or more of the ways shown on the conceptual understanding pentagon below.

Conceptual Understanding Pentagon

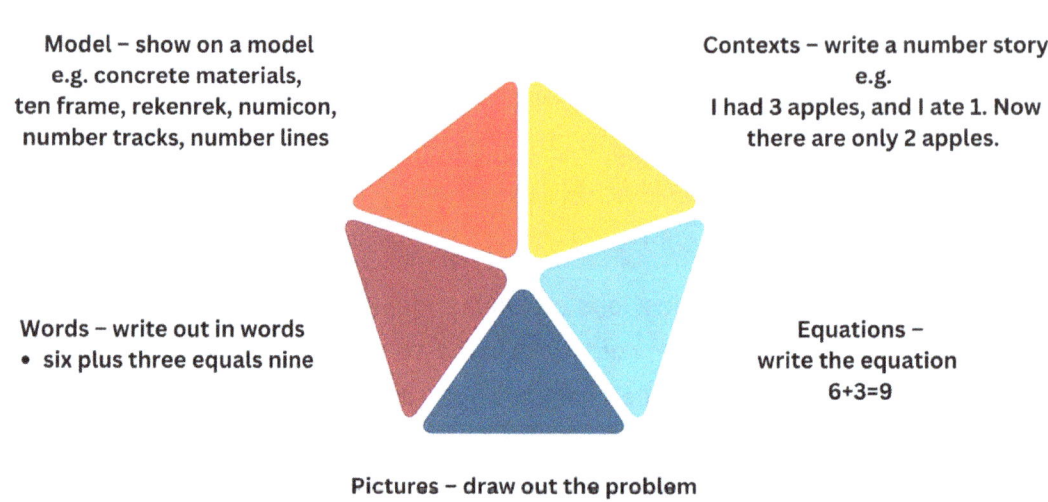

Van de Walle 2006, adapted Mike Flynn

This way students can show their conceptual understanding or misunderstanding:

- Model – show on a model e.g., concrete materials, ten frames, Rekenrek, Numicon, number tracks, number lines.
- Words – write out in works – six plus three equals nine.
- Pictures – draw out the problem as a picture.
- Equations – write the equation (already done for you but you could rewrite vertically or horizontally)
- Contexts – write a number story – e.g. I had 3 apples, and I ate 1. Now there are only 2 apples.

Setting out the student book

$6 + 2 = 8$

Six plus two equals eight.

Jane had 6 puppies and her Poppy brought her 2 more. Now she has 8 puppies!

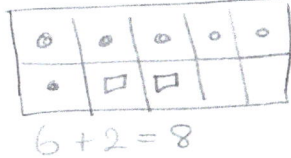

$6 + 2 = 8$

4 - 1 = 3

Four subtract one equals three.

Four birds were in the garden. One flew away. Now there are three birds

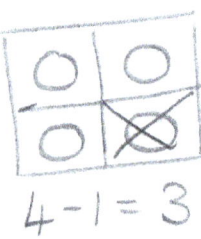

4 - 1 = 3

Coding the answers

Coding the answers will help to diagnose which strategies the students are using and which they are not. We really need students to move beyond counting in ones and seeing numbers in bigger groups as well as learning some of these facts so they can recall them automatically.

Coding the student answer strategy: ask the student how they worked out the answer – they may use more than one way	
Automatic – just knew the answer, immediate recall	A
Counted on or back a small number (+/- 0,1,2,3)	S
Rainbow facts – knew the pair made 10	RF
Counted on from largest number	CO
Counted back for subtraction	CB
Doubled or halved	D or H
Near double	ND
Place value – e.g., 10+4= 14	PV
Compensated – made an adjustment to the number before calculating e.g., 9+6 – rearranged to 10+5 as easier	C
Other – you may wish to note this down	O

Module 5 - Addition

Day	Draw and solve						Code/ Notes
1	5+1=	0 + 0=	1 + 2=	0 + 3=	3 + 1=	4+2=	
2	1 + 5=	2 + 2=	0 + 6=	2 + 3=	5 + 1=	6 + 0=	
3	1 + 1=	0 + 2=	1 + 4=	2 + 4=	4 + 2=	5 + 0=	
4	0 + 1=	2 + 0=	1 + 3=	0 + 4=	4 + 0=	3 + 3=	
5	1 + 0=	0 + 5=	2 + 1=	3 + 0=	3 + 2=	4 + 1=	

Module 6 - Subtraction

Day	Draw and solve						Code/ Notes
1	0 − 0=	2 − 1=	6 − 2=	5 − 1=	6 − 3=	5 − 5=	
2	6 − 0=	4 − 0=	4 − 1=	5 − 2=	2 − 2=	6 − 5=	
3	1 − 1=	1 − 0=	5 − 0=	3 − 2=	6 − 4=	5 − 3=	
4	2 − 0=	3 − 1=	4 − 2=	4 − 4=	5 − 3=	5 − 4=	
5	6 − 1=	3 − 0=	3 − 3=	4 − 3=	6 − 6=	6 − 2=	

Module 7 - Doubling

Day	Draw and solve	Code/ Notes	Retrieval Practice	Code/ Notes
1	0 + 0 = 3 + 3 =		0 + 0= 3 − 3= 5 − 0=	
2	1 + 1= 2 + 2 =		2 − 1= 1 + 0= 6 − 5=	
3	2 + 2 = 3 + 3 =		1 + 1= 2 − 0= 3 + 2=	
4	0 + 0 = 1 + 1=		3 + 0= 5 − 4= 4 + 1=	
5	1 + 1= 2 + 2 =		4 − 2= 6 − 1= 4 + 2=	

Module 8 – Sharing

Day	Draw and solve	Code/ Notes	Draw and solve	Code/ Notes
1	4 share by 2 6 share by 4		5 + 1= 6 – 5= 2 – 0=	
2	5 share by 1 3 share by 3		2 + 1= 5 + 0= 5 – 3=	
3	6 share by 2 4 share by 3		3 + 1= 1 – 1= 4 – 1=	
4	4 share by 4 3 share by 1		4 – 3= 2 – 2= 4 + 0=	
5	2 share by 2 6 share by 3		3 – 2= 6 – 2= 6 + 0=	

Foundation Numbers to 1-6 Word Problem Solving book

How to use these problems

This booklet can be printed from the resource page on the website – see Page 2.

Each of these cards states which problem type is the focus.

The images are there to support but not to be relied upon. Encourage the student to think about what the problem is asking them to do and solve.

The conceptual pentagon provides ideas of how you could go about solving the problem.

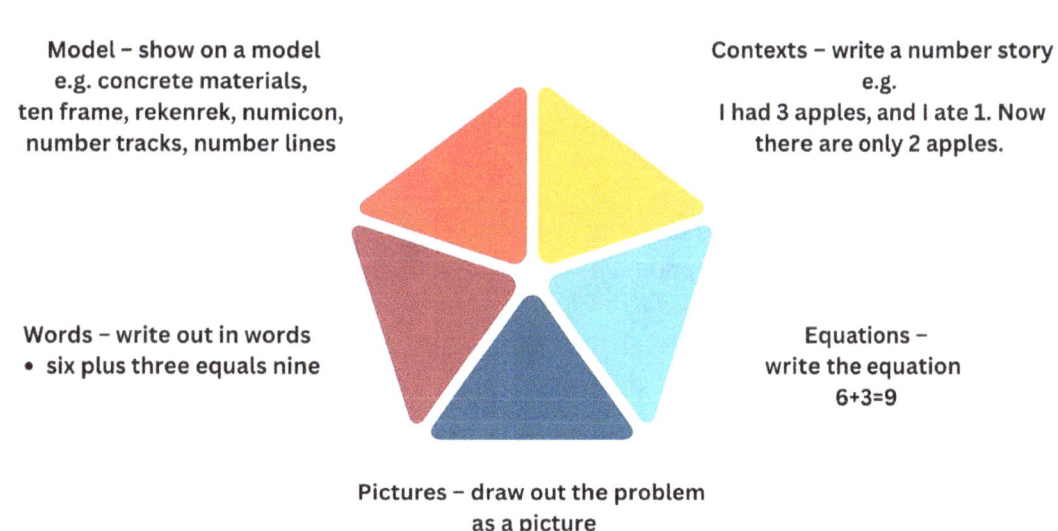

Van de Walle 2006, adapted Mike Flynn

Recording Page

Use this page to record how the student managed with each different problem type. Which can they do easily, and which need more practice?

Used at foundation level:

Join – result unknown 6+2=?	Join – change unknown 6+?=8	Join – start unknown ?+2=8
Separate – result unknown 9-5=?	Separate – change unknown 9-?=4	Separate – start unknown ?-5=4
Part-part-whole – whole unknown 5+4=?	Part-part-whole – part unknown ?+4=9	
Compare – difference unknown 7-2=?	Compare – compared set unknown 7-?=5	Compare – referent unknown ?-5=2

Module 2 – Part part whole - numbers 1-3

Two bunnies are hopping in the garden, and one more bunny joins them. How many bunnies are there now?

Mathtastic Foundation Level
Part Part Whole

There is one apple on the table, and Mum gives you two more apples. How many apples do you have now?

Mathtastic Foundation Level
Part Part Whole

You have two toy cars, and your friend gives you one more car to play with. How many toy cars do you have in total?

Mathtastic Foundation Level
Part Part Whole

There is one puppy playing in the yard. Two more puppies join. How many puppies are playing now?

Mathtastic Foundation Level
Part Part Whole

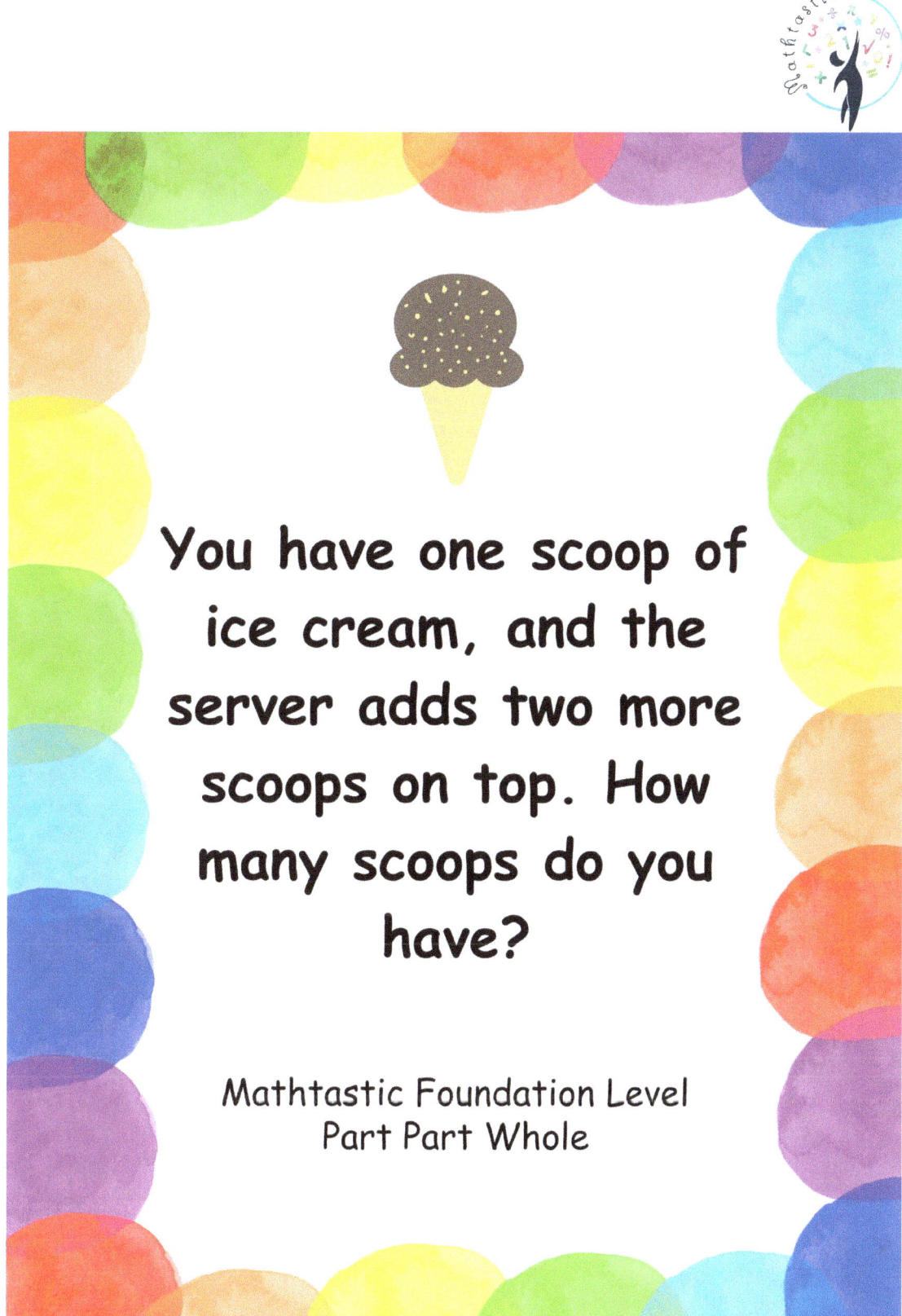

You have one scoop of ice cream, and the server adds two more scoops on top. How many scoops do you have?

Mathtastic Foundation Level
Part Part Whole

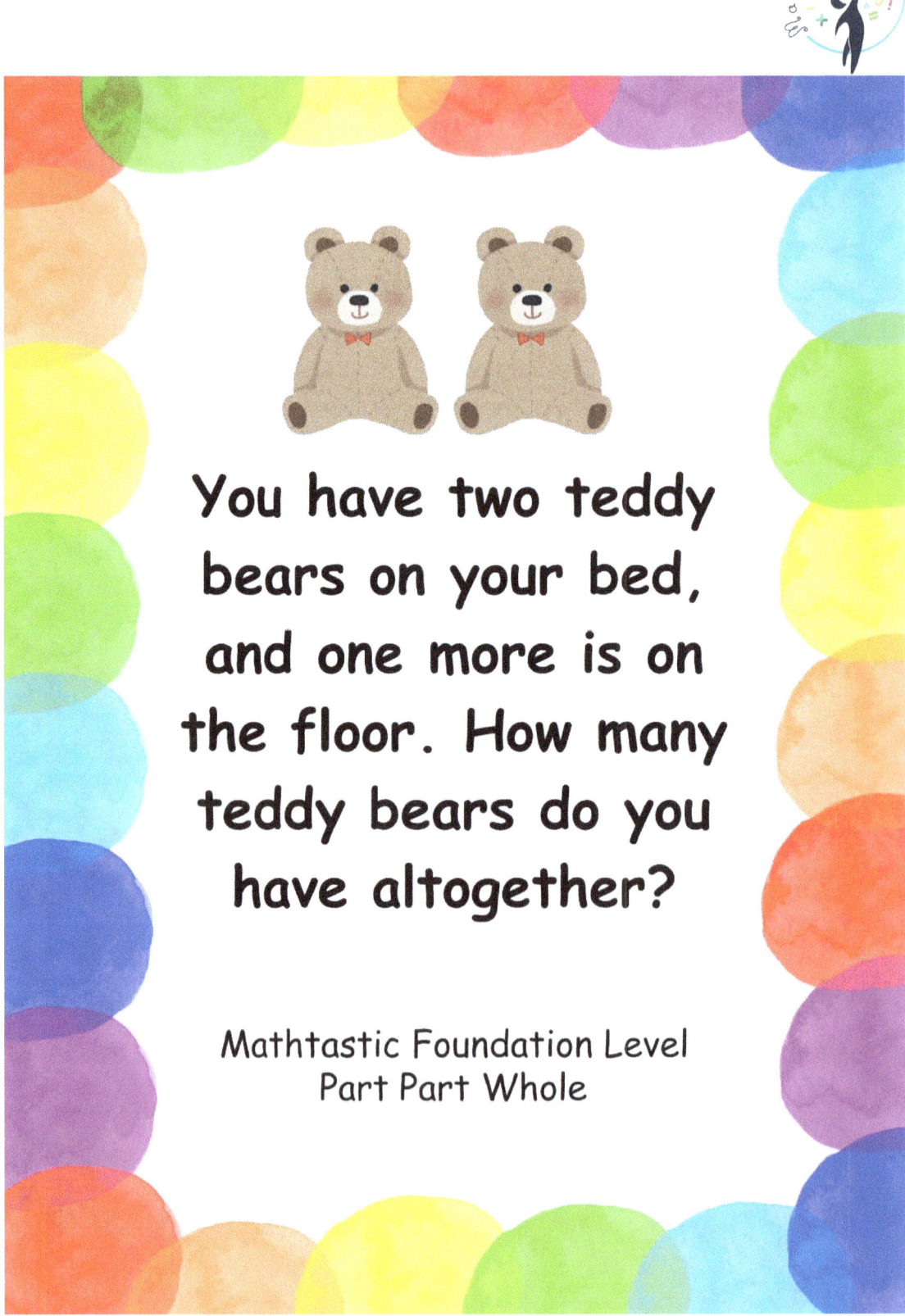

You have two teddy bears on your bed, and one more is on the floor. How many teddy bears do you have altogether?

Mathtastic Foundation Level
Part Part Whole

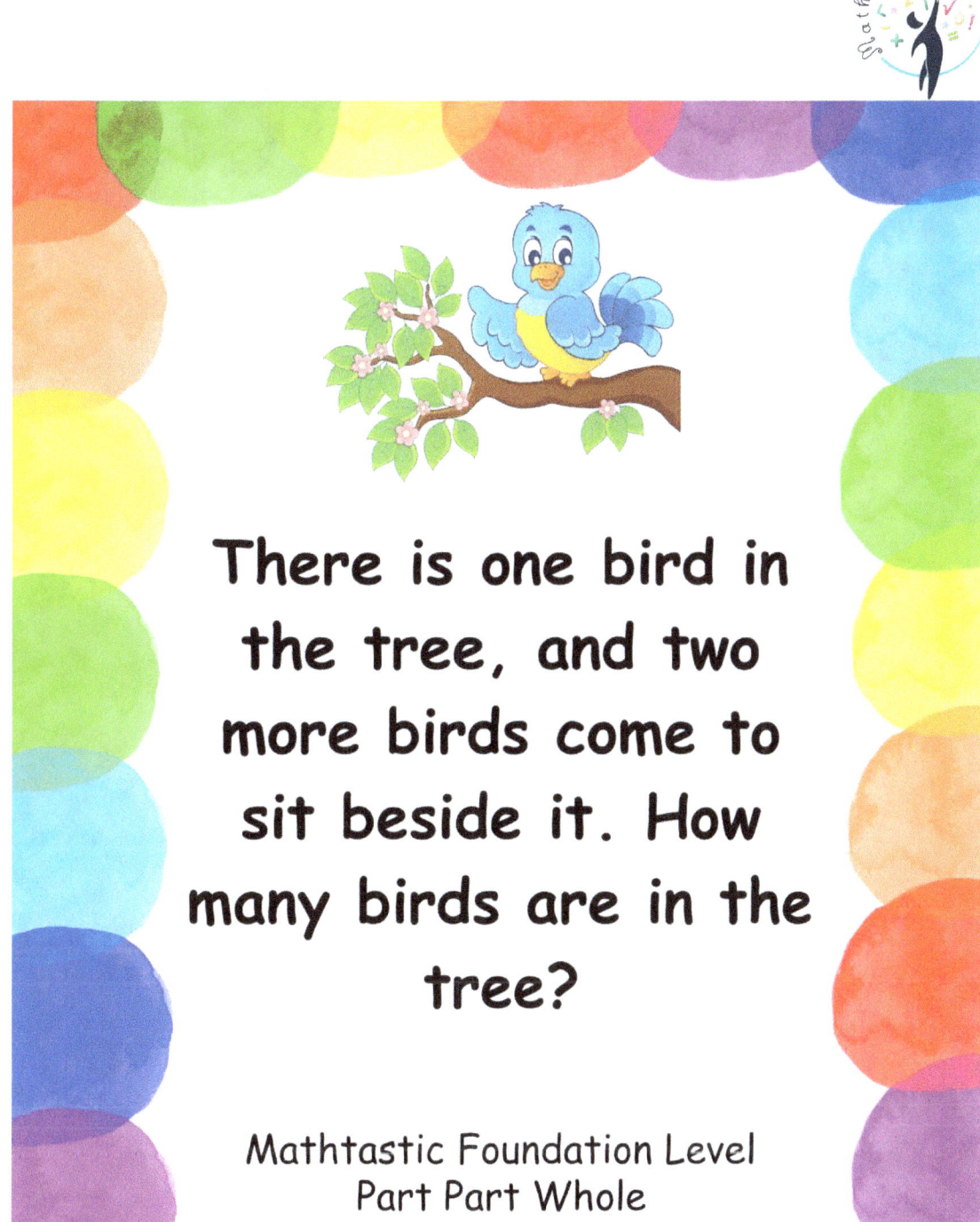

There is one bird in the tree, and two more birds come to sit beside it. How many birds are in the tree?

Mathtastic Foundation Level
Part Part Whole

You have two cookies, and Mum gives you one more cookie. How many cookies do you have now?

Mathtastic Foundation Level
Part Part Whole

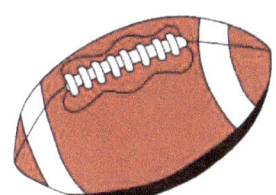

You have one red ball and two blue balls. How many balls do you have to play with?

Mathtastic Foundation Level
Part Part Whole

Module 2 – Addition by joining - numbers 1-3

One small dinosaur is walking through the forest. Two more dinosaurs join it. How many dinosaurs are walking together now?

Mathtastic Foundation Level
Joining

A train has two cars attached. One more car joins the train. How many train cars are there now?

Mathtastic Foundation Level
Joining

You are holding one marble. Your friend gives you two more marbles to play with. How many marbles do you have now?

Mathtastic Foundation Level
Joining

You start with two snowballs to make a snowman. You add one more snowball. How many snowballs did you use to build the snowman?

Mathtastic Foundation Level
Joining

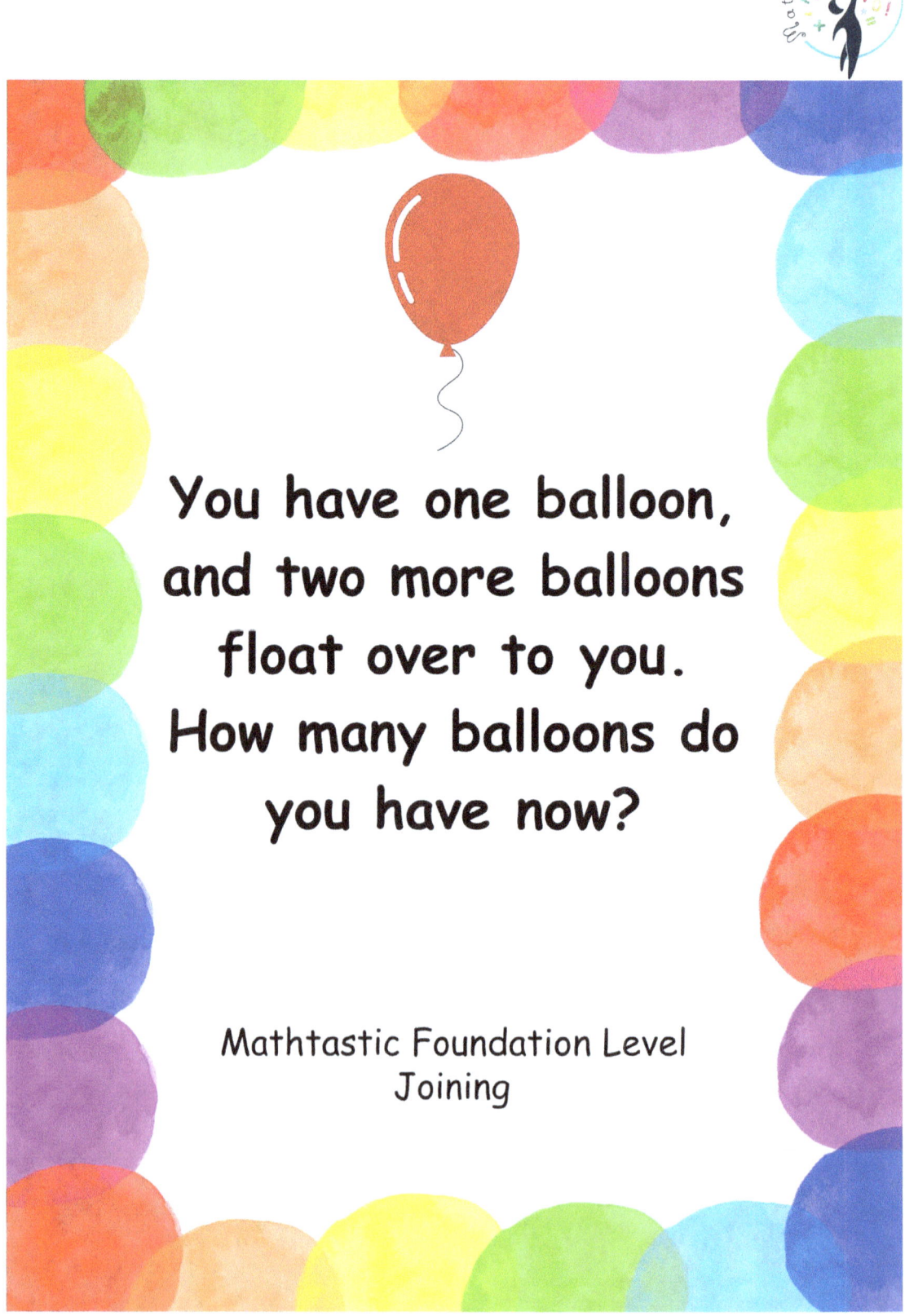

You have one balloon, and two more balloons float over to you. How many balloons do you have now?

Mathtastic Foundation Level
Joining

One butterfly is sitting on a flower. Two more butterflies join it. How many butterflies are on the flower now?

Mathtastic Foundation Level
Joining

You have two shells in your bucket. You pick up one more shell. How many shells are in your bucket now?

Mathtastic Foundation Level
Joining

One cat is napping in the sun. Two more cats come to join it. How many cats are napping now?

Mathtastic Foundation Level
Joining

You pick one flower from the garden. Then you pick two more flowers. How many flowers do you have now?

Mathtastic Foundation Level
Joining

There are two fish swimming in the pond. One more fish joins them. How many fish are swimming together?

Mathtastic Foundation Level
Joining

Module 3 - Remove/ subtract - numbers 1-3

You have three pencils. You give one pencil to your friend. How many pencils do you have left?

Mathtastic Foundation Level
Remove/Subtract

There are two birds sitting in a tree. One bird flies away. How many birds are left in the tree?

Mathtastic Foundation Level
Remove/Subtract

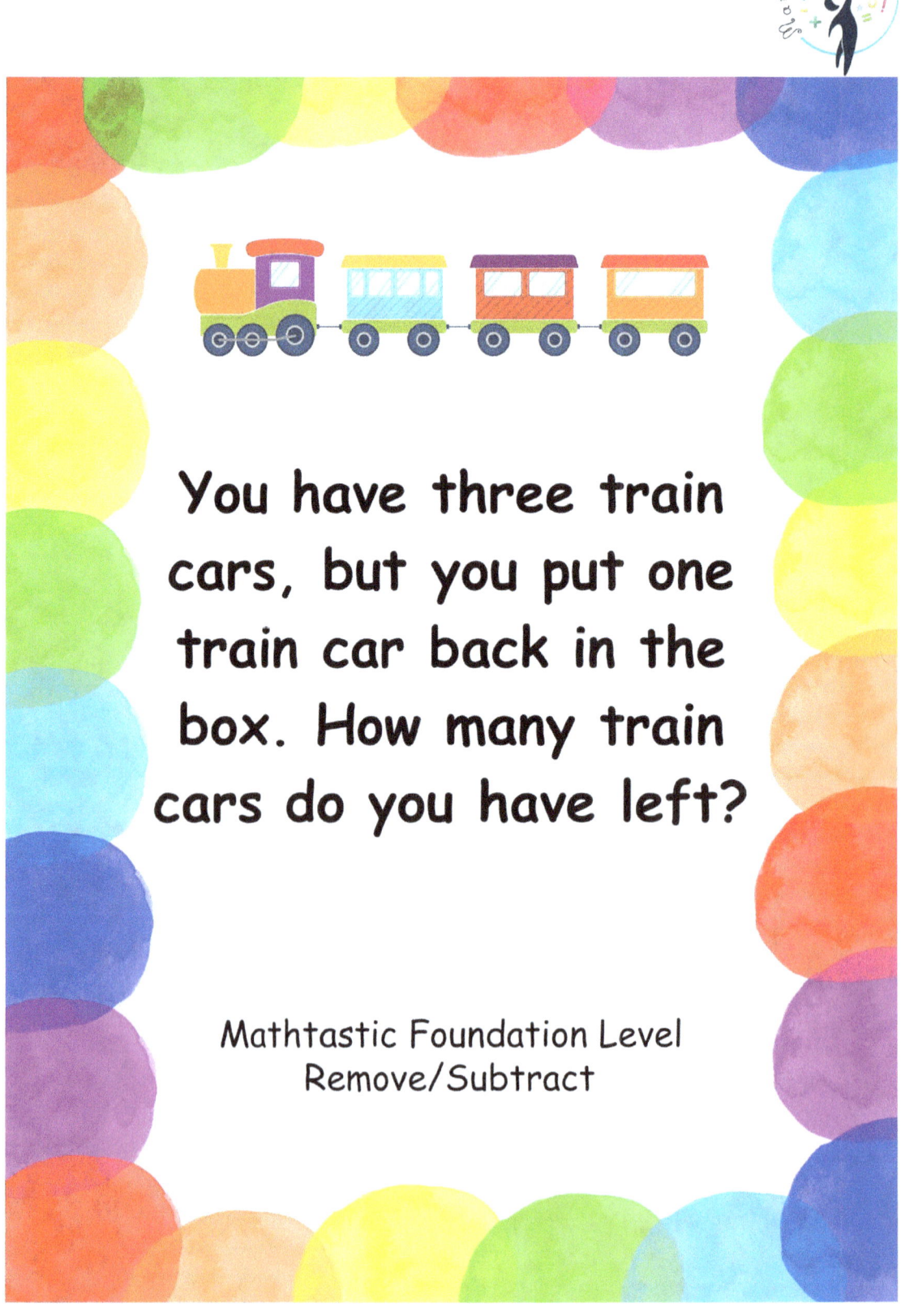

You have three train cars, but you put one train car back in the box. How many train cars do you have left?

Mathtastic Foundation Level
Remove/Subtract

You are playing with three teddy bears. You put two teddy bears away. How many teddy bears are still out?

Mathtastic Foundation Level
Remove/Subtract

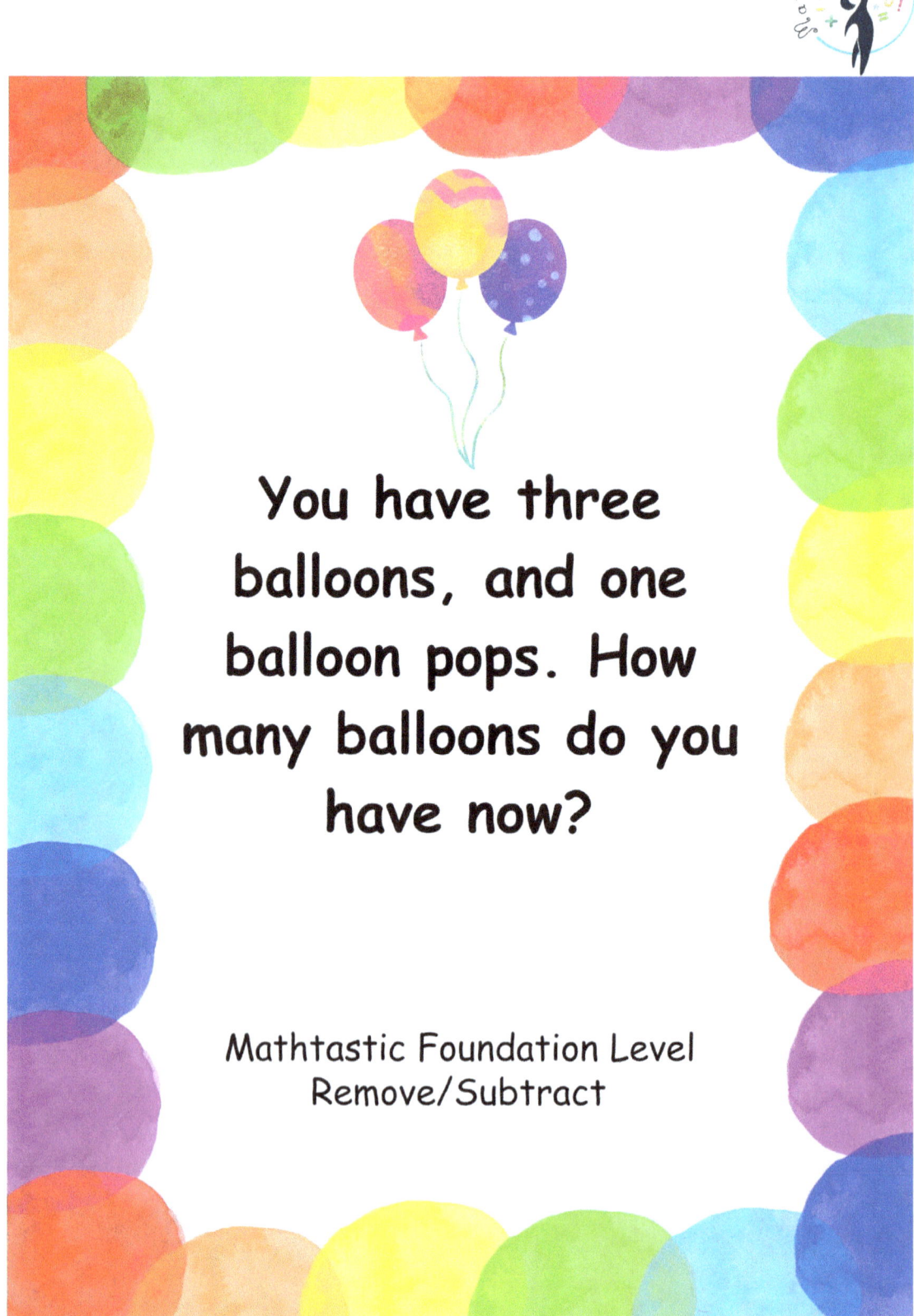

You have three balloons, and one balloon pops. How many balloons do you have now?

Mathtastic Foundation Level
Remove/Subtract

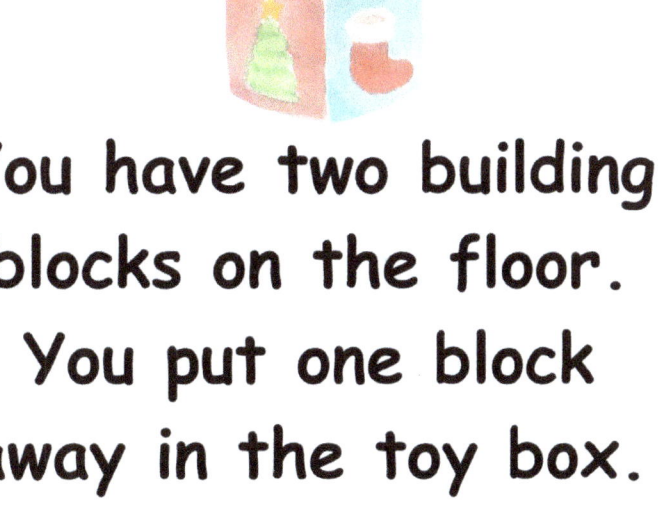

You have two building blocks on the floor. You put one block away in the toy box. How many blocks are left on the floor?

Mathtastic Foundation Level
Remove/Subtract

You have three apples on the table. You eat one apple. How many apples are left on the table?

Mathtastic Foundation Level
Remove/Subtract

You have three shells. You give two shells to your friend. How many shells do you have now?

Mathtastic Foundation Level
Remove/Subtract

There are three cookies in the jar. You take one cookie out to eat. How many cookies are left in the jar?

Mathtastic Foundation Level
Remove/Subtract

You have two toy cars on the table. You put one car back in your toy box. How many toy cars are left on the table?

Mathtastic Foundation Level
Remove/Subtract

Module 4 - Addition or Subtraction - numbers 1-3

There are 2 birds on a tree. 1 more bird joins them. How many birds are on the tree now?

Mathtastic Foundation Level
Addition or Subtraction
Numbers 1-3

Anna has 3 lollies. She eats 1 lolly. How many lollies does Anna have left?

Mathtastic Foundation Level
Addition or Subtraction
Numbers 1-3

Emma has 2 cookies. She bakes 1 more cookie. How many cookies does Emma have now?

Mathtastic Foundation Level
Addition or Subtraction
Numbers 1-3

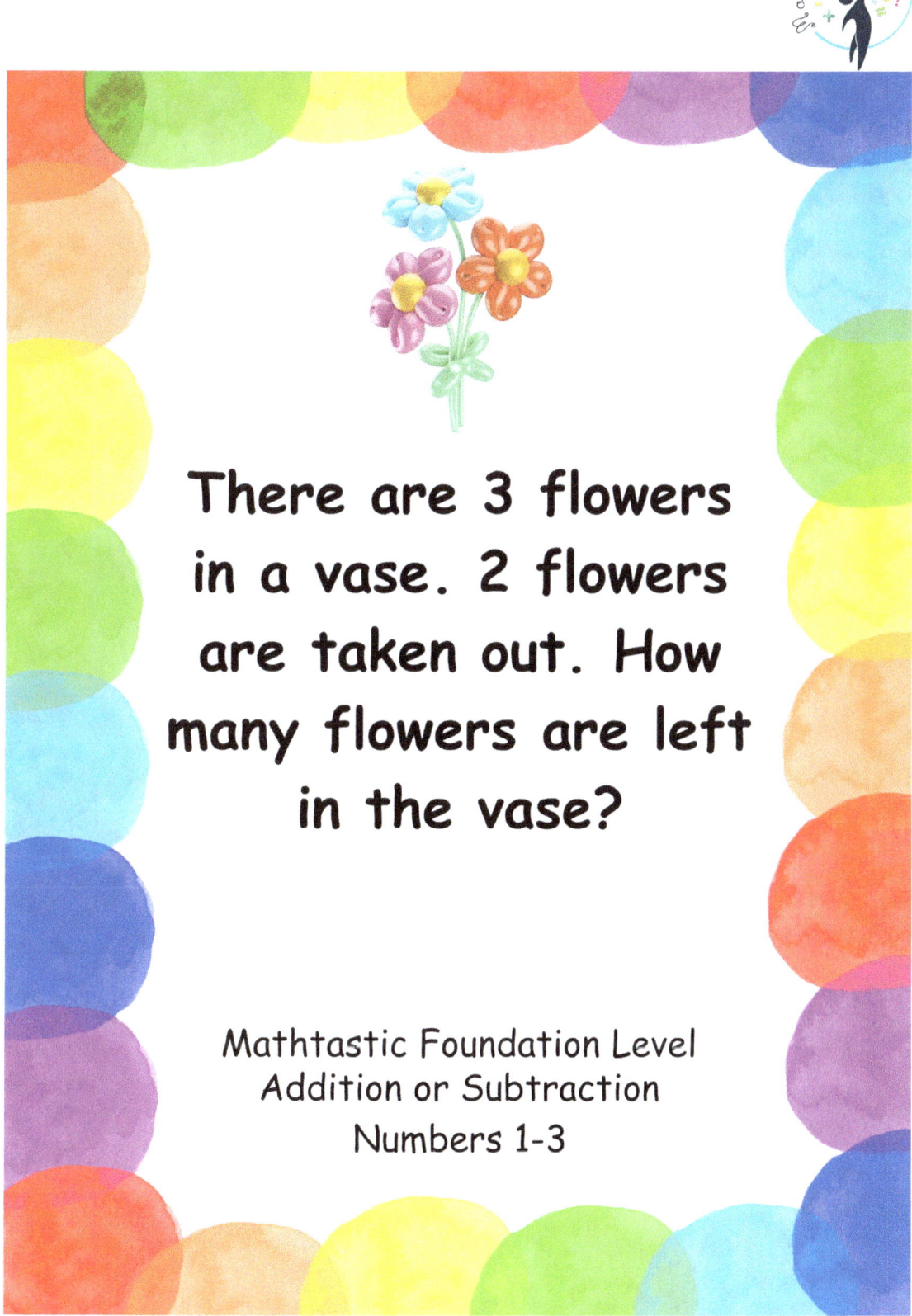

There are 3 flowers in a vase. 2 flowers are taken out. How many flowers are left in the vase?

Mathtastic Foundation Level
Addition or Subtraction
Numbers 1-3

Tom has 1 apple. He finds 2 more apples. How many apples does Tom have now?

Mathtastic Foundation Level
Addition or Subtraction
Numbers 1-3

There are 2 ducks in a pond. 1 duck flies away. How many ducks are left in the pond?

Mathtastic Foundation Level
Addition or Subtraction
Numbers 1-3

Liam has 1 toy car. He gets 2 more toy cars for his birthday. How many toy cars does Liam have now?

Mathtastic Foundation Level
Addition or Subtraction
Numbers 1-3

Mia has 2 crayons.
She loses 1 crayon.
How many crayons
does Mia have now?

Mathtastic Foundation Level
Addition or Subtraction
Numbers 1-3

Ben has 3 toy cars. He gives 1 toy car to his friend. How many toy cars does Ben have now?

Mathtastic Foundation Level
Addition or Subtraction
Numbers 1-3

There are 3 apples on a table. 1 apple is eaten. How many apples are left on the table?

Mathtastic Foundation Level
Addition or Subtraction
Numbers 1-3

Sara has 2 balloons. Her friend gives her 1 more balloon. How many balloons does Sara have now?

Mathtastic Foundation Level
Addition or Subtraction
Numbers 1-3

There are 2 frogs in a pond. 1 more frog jumps in. How many frogs are in the pond now?

Mathtastic Foundation Level
Addition or Subtraction
Numbers 1-3

Module 5 - Addition part part whole - numbers 1-6

You have 2 red blocks and 3 blue blocks. How many blocks do you have in total?

Mathtastic Foundation Level
Part Part Whole
Numbers 1-6

There are 1 cat and 4 dogs in the yard. How many animals are in the yard?

Mathtastic Foundation Level
Part Part Whole
Numbers 1-6

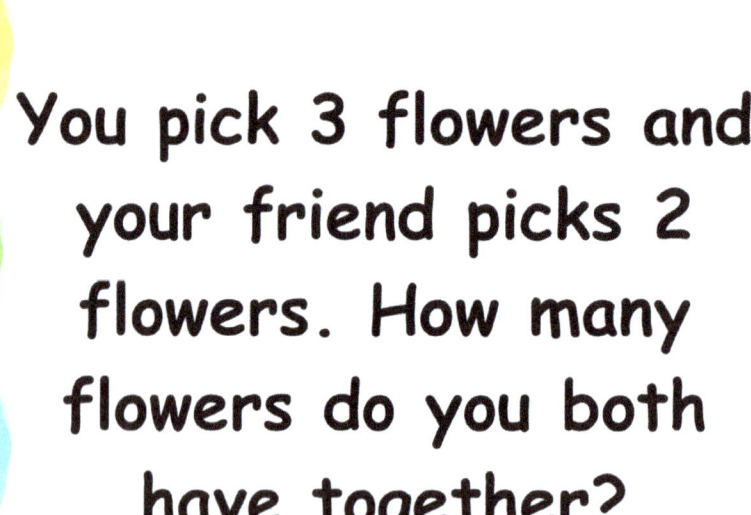

You pick 3 flowers and your friend picks 2 flowers. How many flowers do you both have together?

Mathtastic Foundation Level
Part Part Whole
Numbers 1-6

There are 5 apples in one basket and 1 apple in another basket. How many apples are there in total?

Mathtastic Foundation Level
Part Part Whole
Numbers 1-6

You have 4 toy cars and your brother has 2 toy cars. How many toy cars do you both have?

Mathtastic Foundation Level
Part Part Whole
Numbers 1-6

There are 3 birds on one tree and 3 birds on another tree. How many birds are there in total?

Mathtastic Foundation Level
Part Part Whole
Numbers 1-6

You have 1 gingerbread man and your friend has 5. How many gingerbread men do you have altogether?

Mathtastic Foundation Level
Part Part Whole
Numbers 1-6

There are 2 frogs in one pond and 4 frogs in another pond. How many frogs are there in total?

Mathtastic Foundation Level
Part Part Whole
Numbers 1-6

You have 3 crayons and your sister has 1 crayon. How many crayons do you have?

Mathtastic Foundation Level
Part Part Whole
Numbers 1-6

There are 4 ducks in one pond and 1 duck in another pond. How many ducks are there in total?

Mathtastic Foundation Level
Part Part Whole
Numbers 1-6

You have 3 balloons and your friend has 2 more balloons. How many balloons do you have altogether?

Mathtastic Foundation Level
Part Part Whole
Numbers 1-6

There are 1 fish in one tank and 2 fish in another tank. How many fish are there in total?

Mathtastic Foundation Level
Part Part Whole
Numbers 1-6

You have 4 marbles and your friend has 1 more marble. How many marbles are there?

Mathtastic Foundation Level
Part Part Whole
Numbers 1-6

There are 2 apples in one basket and 2 apples in another basket. How many apples are there in total?

Mathtastic Foundation Level
Part Part Whole
Numbers 1-6

You have 4 pencils and your friend gives you 2 more pencils. How many pencils do you have now?

Mathtastic Foundation Level
Part Part Whole
Numbers 1-6

Module 5 - Addition by joining – numbers 1-6

You have 1 toy car. Your friend gives you 2 more toy cars. How many toy cars do you have now?

Mathtastic Foundation Level
Joining
Numbers 1-6

There are 3 birds in a nest. 2 more birds join them. How many birds are there now?

Mathtastic Foundation Level
Joining
Numbers 1-6

You have 2 cherries. Your mom gives you 3 more cherries. How many cherries do you have now?

Mathtastic Foundation Level
Joining
Numbers 1-6

There are 4 ducks in a line. 1 more duck joins them. How many ducks are there now?

Mathtastic Foundation Level
Joining
Numbers 1-6

You have 1 balloon. Your friend gives you 4 more balloons. How many balloons do you have now?

Mathtastic Foundation Level
Joining
Numbers 1-6

There are 2 frogs on a lily pad. 3 more frogs join them. How many frogs are there now?

Mathtastic Foundation Level
Joining
Numbers 1-6

You have 3 crayons. Your sister gives you 2 more crayons. How many crayons do you have now?

Mathtastic Foundation Level
Joining
Numbers 1-6

There are 5 flowers in the garden. 1 more flower blooms. How many flowers are there now?

Mathtastic Foundation Level
Joining
Numbers 1-6

You have 2 cookies. Your friend gives you 2 more cookies. How many cookies do you have now?

Mathtastic Foundation Level
Joining
Numbers 1-6

There is 1 cat in the yard. 3 more cats join it. How many cats are there now?

Mathtastic Foundation Level
Joining
Numbers 1-6

You have 4 marbles. Your friend gives you 1 more marble. How many marbles do you have now?

Mathtastic Foundation Level
Joining
Numbers 1-6

There are 3 fish in the tank. 2 more fish are added. How many fish are there now?

Mathtastic Foundation Level
Joining
Numbers 1-6

You have 2 pencils. Your teacher gives you 3 more pencils. How many pencils do you have now?

Mathtastic Foundation Level
Joining
Numbers 1-6

There are 1 dog in the park. 4 more dogs join it. How many dogs are there now?

Mathtastic Foundation Level
Joining
Numbers 1-6

You have 5 blocks. Your friend gives you 1 more block. How many blocks do you have now?

Mathtastic Foundation Level
Joining
Numbers 1-6

Module 6 - Subtraction - numbers 1-6

You have 5 apples. You eat 2 apples. How many apples do you have left?

Mathtastic Foundation Level
Subtraction
Numbers 1-6

There are 4 birds on a tree. 1 bird flies away. How many birds are left on the tree?

Mathtastic Foundation Level
Subtraction
Numbers 1-6

You have 3 balloons. 1 balloon pops. How many balloons do you have now?

Mathtastic Foundation Level
Subtraction
Numbers 1-6

There are 6 ducks in a pond. 3 ducks swim away. How many ducks are left in the pond?

Mathtastic Foundation Level
Subtraction
Numbers 1-6

You have 2 cookies. You give 1 cookie to your friend. How many cookies do you have left?

Mathtastic Foundation Level
Subtraction
Numbers 1-6

There are 5 frogs on a lily pad. 2 frogs jump off. How many frogs are left on the lily pad?

Mathtastic Foundation Level
Subtraction
Numbers 1-6

You have 4 toy cars. You give 2 toy cars to your brother. How many toy cars do you have now?

Mathtastic Foundation Level
Subtraction
Numbers 1-6

There are 3 flowers in a vase. 1 flower is taken out. How many flowers are left in the vase?

Mathtastic Foundation Level
Subtraction
Numbers 1-6

You have 6 crayons.
You lose 3 crayons.
How many crayons do you have now?

Mathtastic Foundation Level
Subtraction
Numbers 1-6

There are 4 cats in the yard. 1 cat goes inside. How many cats are left in the yard?

Mathtastic Foundation Level
Subtraction
Numbers 1-6

You have 4 marbles. You give 1 marble to your friend. How many marbles do you have now?

Mathtastic Foundation Level
Subtraction
Numbers 1-6

There are 4 fish in a tank. 3 fish are moved to another tank. How many fish are left in the tank?

Mathtastic Foundation Level
Subtraction
Numbers 1-6

You have 5 pencils. You give 1 pencil to your sister. How many pencils do you have now?

Mathtastic Foundation Level
Subtraction
Numbers 1-6

There are 6 blocks. You use 2 blocks to build a tower. How many blocks are left?

Mathtastic Foundation Level
Subtraction
Numbers 1-6

You light 2 candles
You blow out 1 candle.
How many candles are left?

Mathtastic Foundation Level
Subtraction
Numbers 1-6

Module 7 - Doubling - numbers 1-6

You have 0 apples. If you double the number of apples, how many apples do you have?

Mathtastic Foundation Level
Doubling
Numbers 1-6

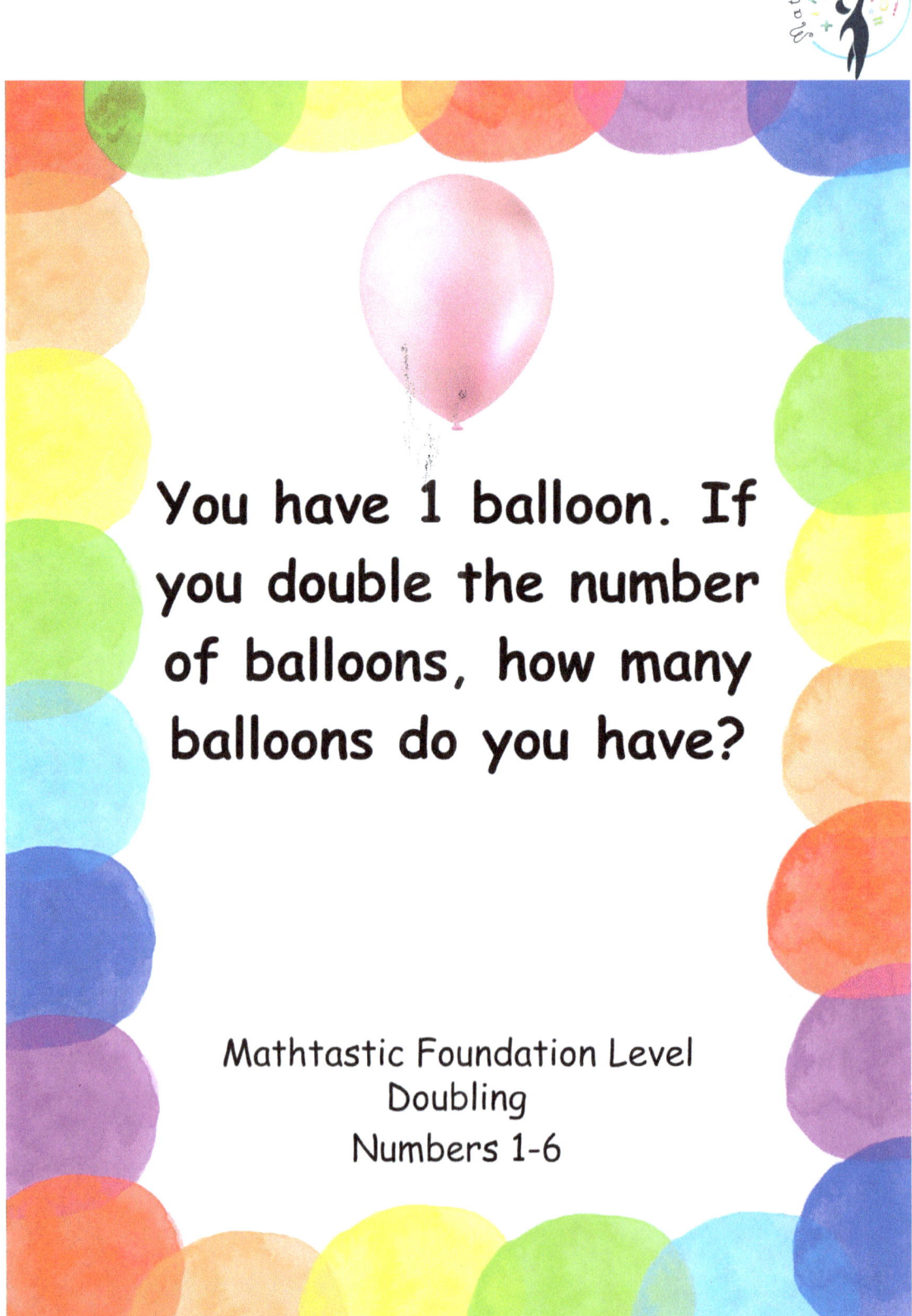

You have 1 balloon. If you double the number of balloons, how many balloons do you have?

Mathtastic Foundation Level
Doubling
Numbers 1-6

You have 2 cookies. If you double the number of cookies, how many cookies do you have?

Mathtastic Foundation Level
Doubling
Numbers 1-6

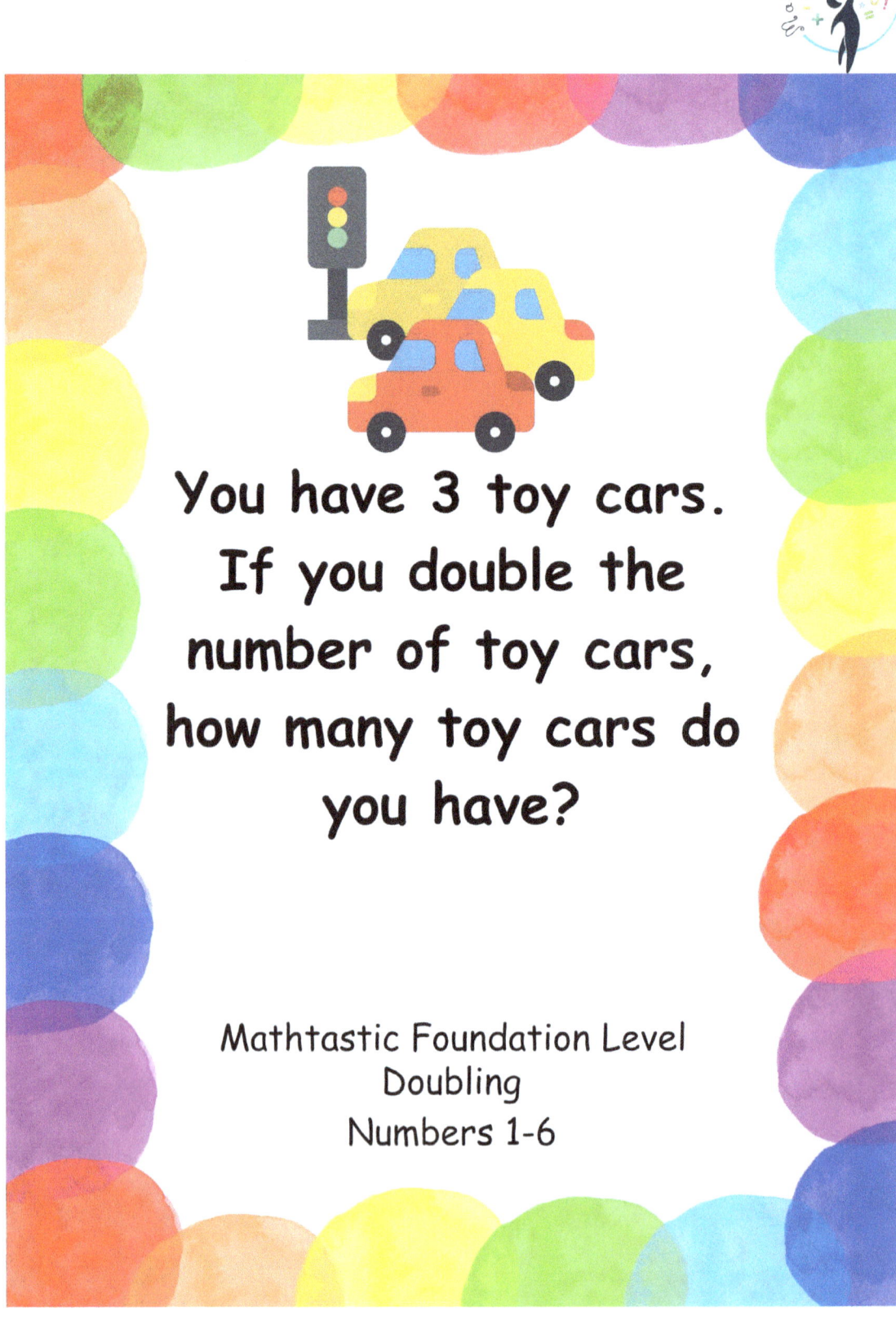

You have 3 toy cars. If you double the number of toy cars, how many toy cars do you have?

Mathtastic Foundation Level
Doubling
Numbers 1-6

You have 0 pencils. If you double the number of pencils, how many pencils do you have?

Mathtastic Foundation Level
Doubling
Numbers 1-6

You have 1 frog. If you double the number of frogs, how many frogs do you have?

Mathtastic Foundation Level
Doubling
Numbers 1-6

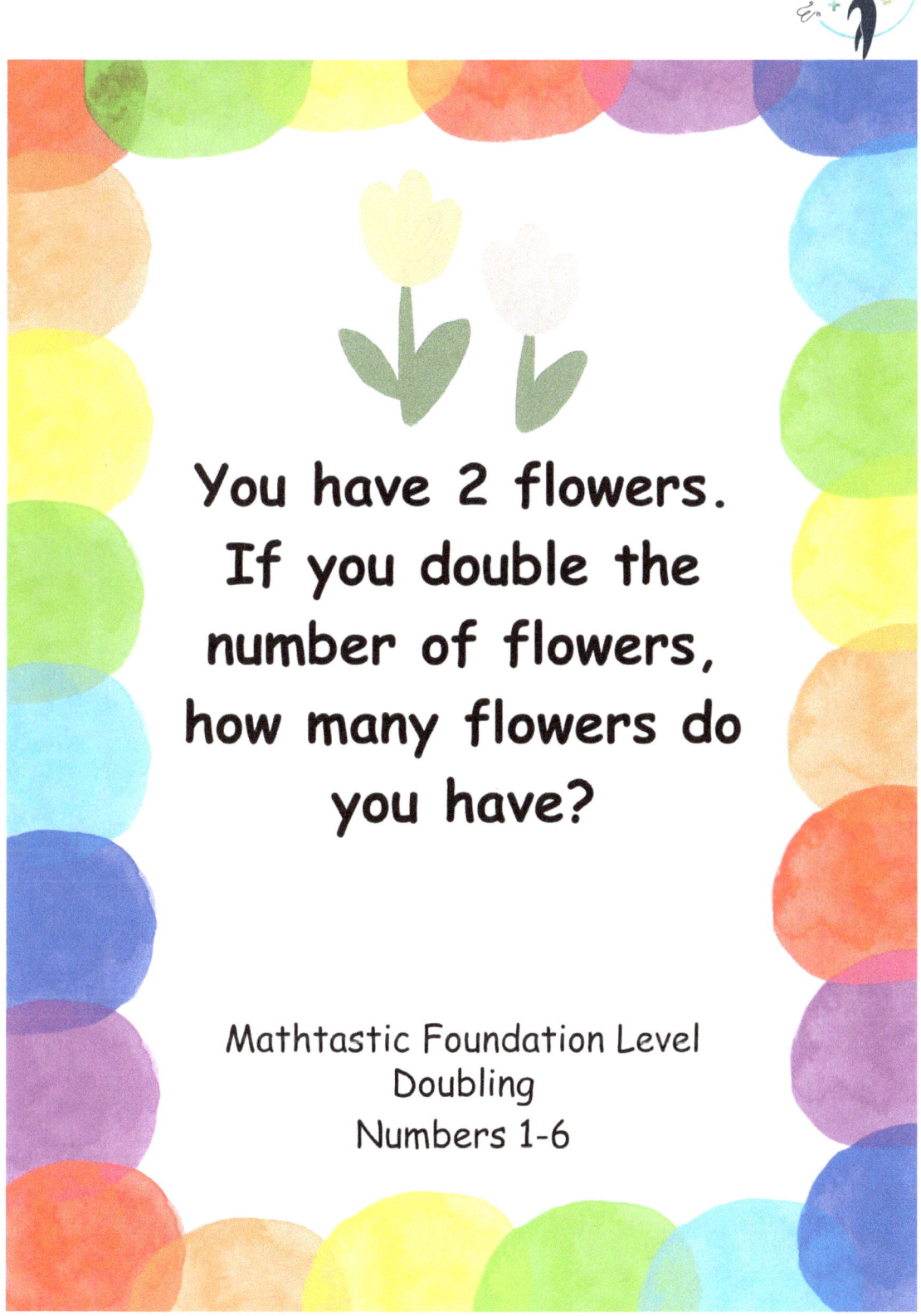

You have 2 flowers. If you double the number of flowers, how many flowers do you have?

Mathtastic Foundation Level
Doubling
Numbers 1-6

You have 3 marbles. If you double the number of marbles, how many marbles do you have?

Mathtastic Foundation Level
Doubling
Numbers 1-6

You have 0 lollies. If you double the number of lollies, how many lollies do you have?

Mathtastic Foundation Level
Doubling
Numbers 1-6

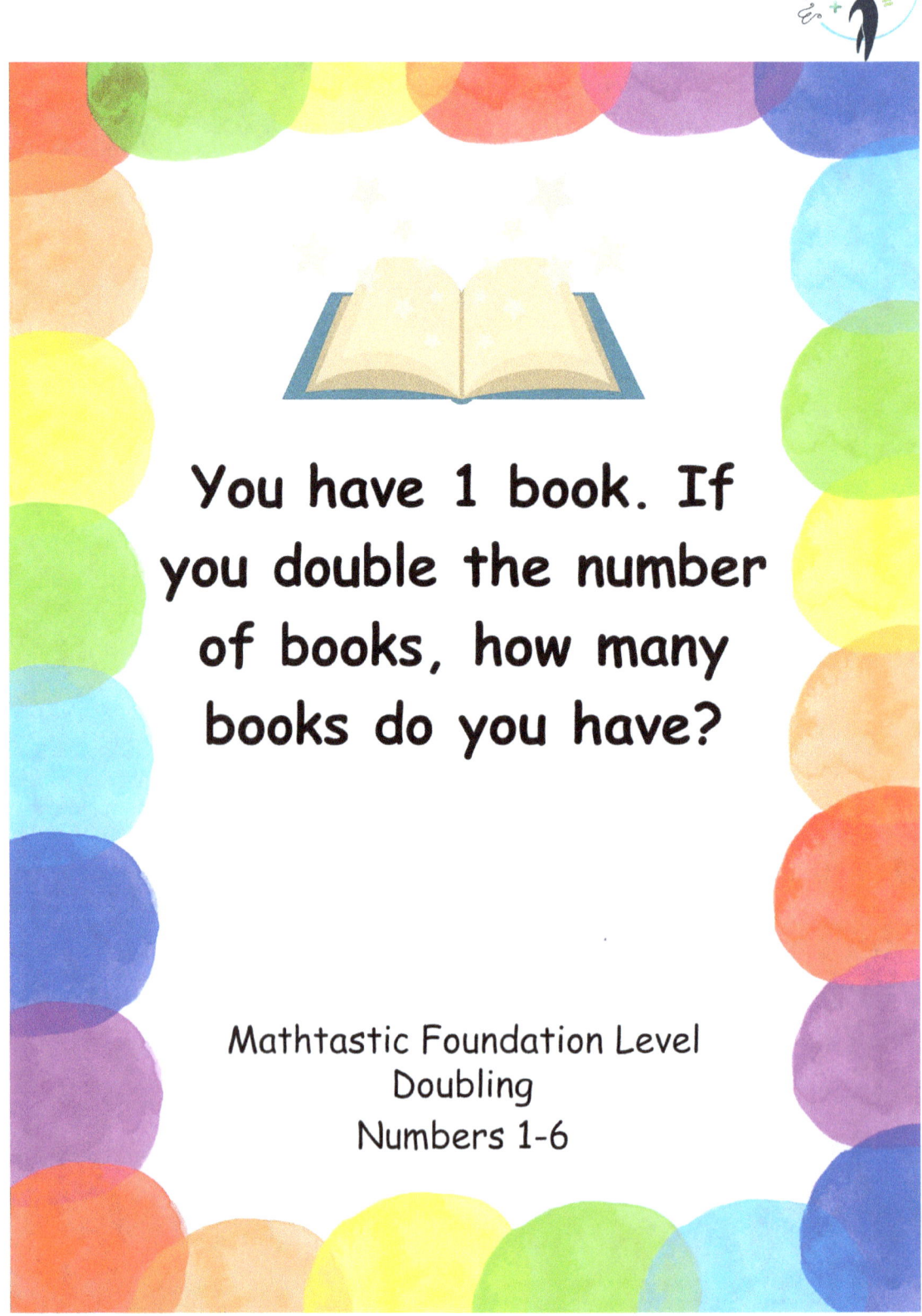

You have 1 book. If you double the number of books, how many books do you have?

Mathtastic Foundation Level
Doubling
Numbers 1-6

You have 2 crayons. If you double the number of crayons, how many crayons do you have?

Mathtastic Foundation Level
Doubling
Numbers 1-6

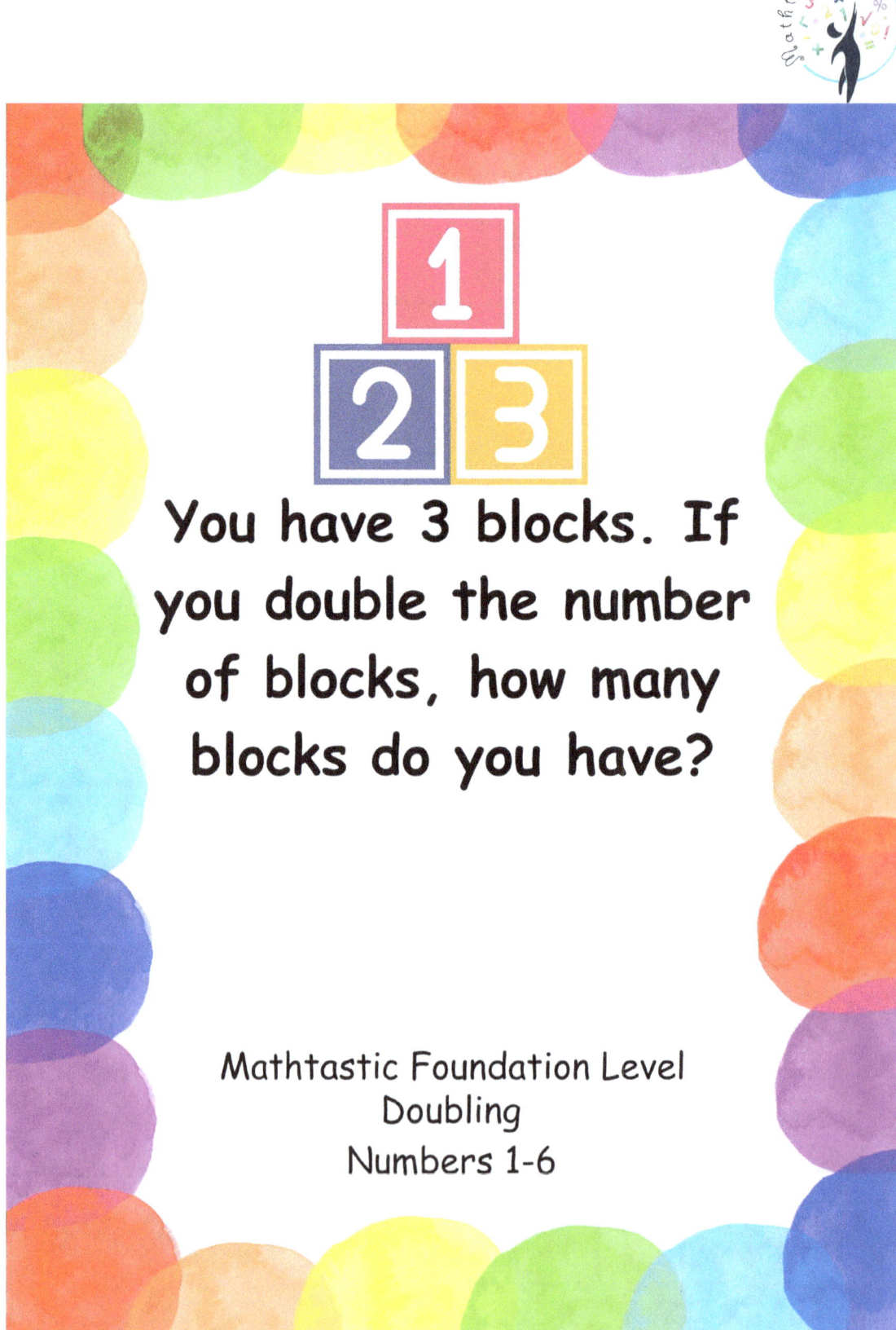

You have 3 blocks. If you double the number of blocks, how many blocks do you have?

Mathtastic Foundation Level
Doubling
Numbers 1-6

You have 0 stickers. If you double the number of stickers, how many stickers do you have?

Mathtastic Foundation Level
Doubling
Numbers 1-6

You have 1 toy. If you double the number of toys, how many toys do you have?

Mathtastic Foundation Level
Doubling
Numbers 1-6

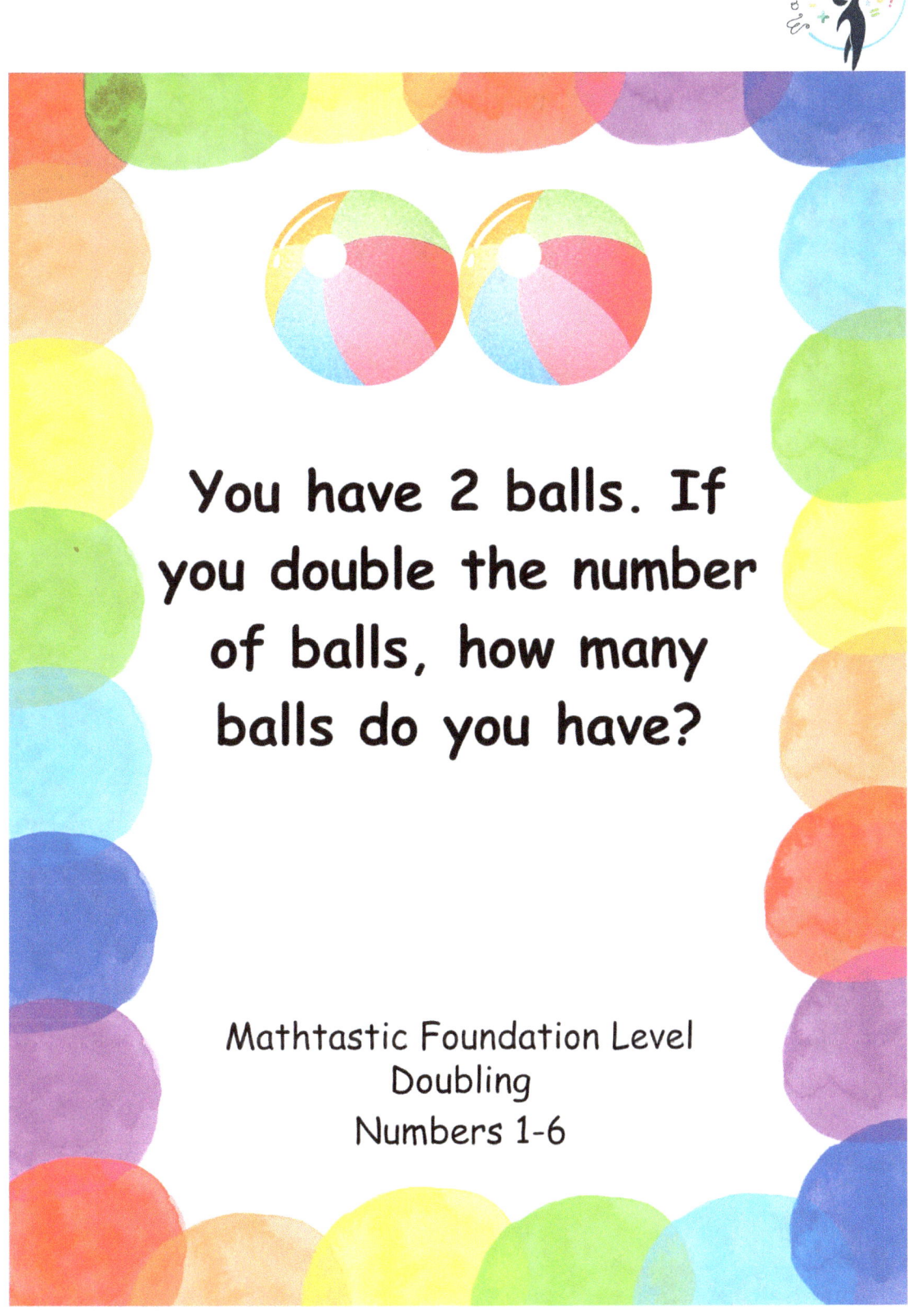

You have 2 balls. If you double the number of balls, how many balls do you have?

Mathtastic Foundation Level
Doubling
Numbers 1-6

Module 8 - Sharing - numbers 1-6

You have 6 apples. You want to share them equally with 2 friends. How many apples does each friend get?

Mathtastic Foundation Level
Sharing
Numbers 1-6

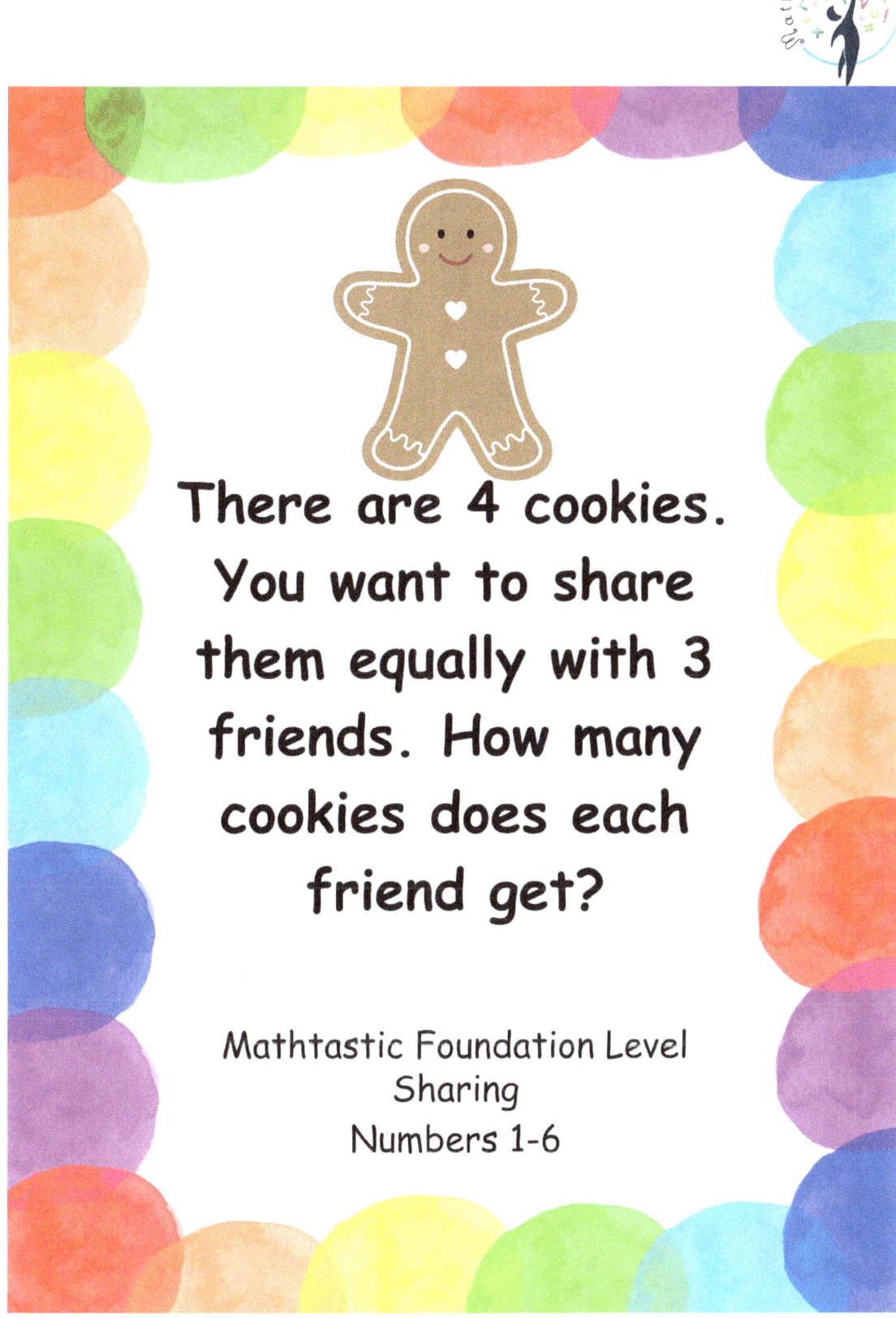

There are 4 cookies. You want to share them equally with 3 friends. How many cookies does each friend get?

Mathtastic Foundation Level
Sharing
Numbers 1-6

You have 5 balloons. You want to share them equally with 2 friends. How many balloons does each friend get?

Mathtastic Foundation Level
Sharing
Numbers 1-6

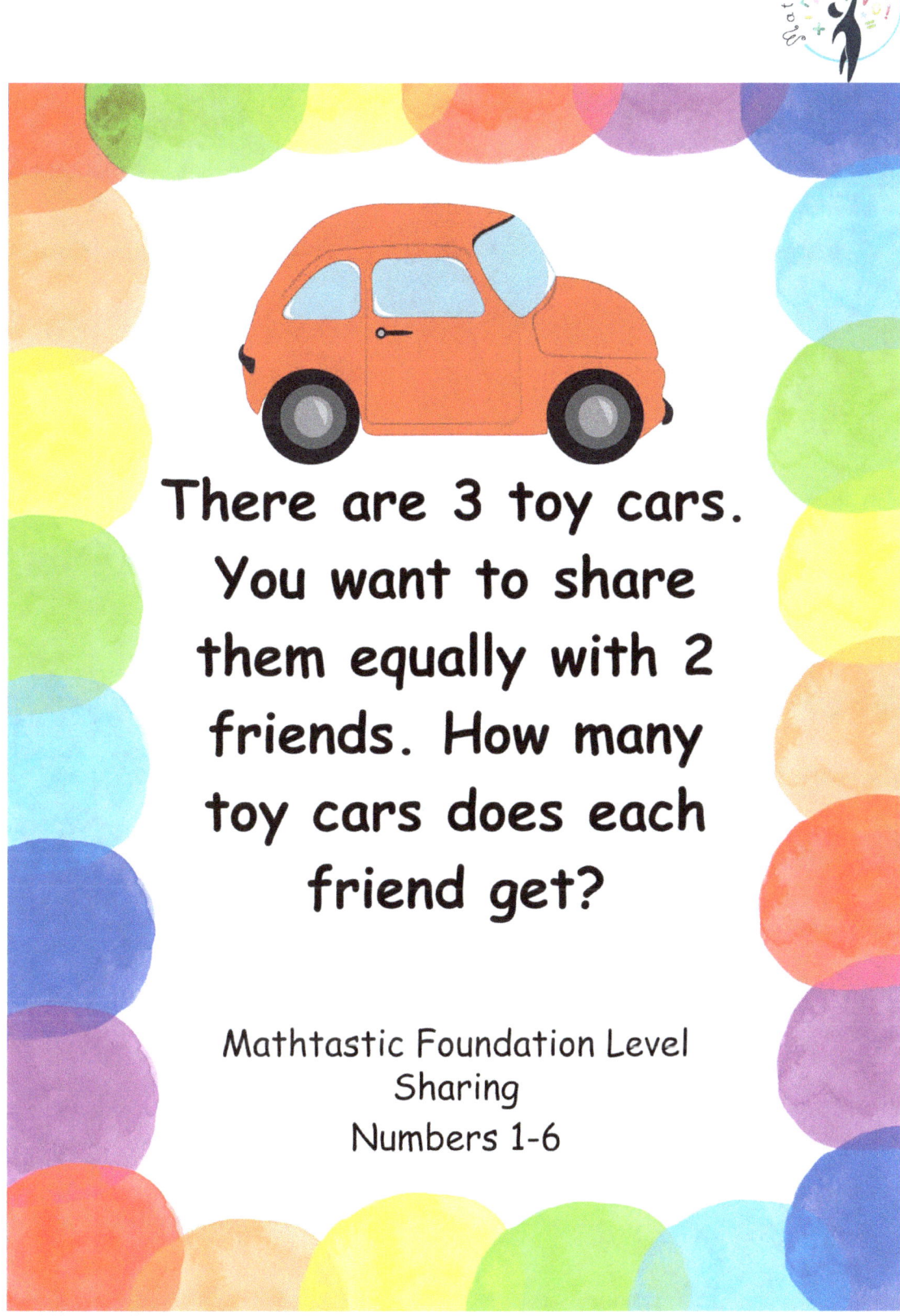

There are 3 toy cars. You want to share them equally with 2 friends. How many toy cars does each friend get?

Mathtastic Foundation Level
Sharing
Numbers 1-6

You have 6 lollipops. You want to share them equally with 4 friends. How many lollipops does each friend get?

Mathtastic Foundation Level
Sharing
Numbers 1-6

There are 2 pencils. You want to share them equally with 3 friends. How many pencils does each friend get?

Mathtastic Foundation Level
Sharing
Numbers 1-6

You have 4 marbles. You want to share them equally with 2 friends. How many marbles does each friend get?

Mathtastic Foundation Level
Sharing
Numbers 1-6

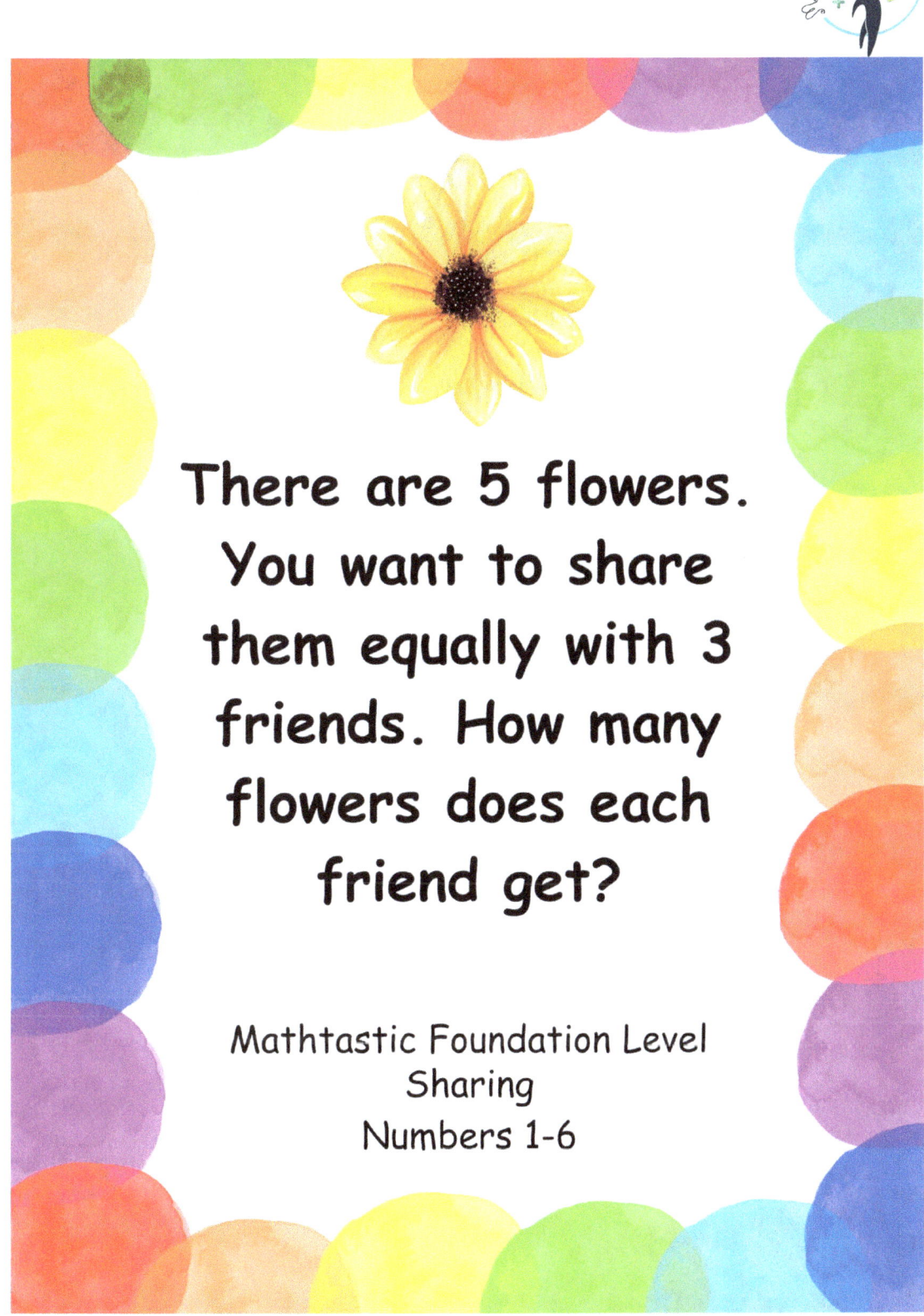

There are 5 flowers. You want to share them equally with 3 friends. How many flowers does each friend get?

Mathtastic Foundation Level
Sharing
Numbers 1-6

You have 6 stickers. You want to share them equally with 5 friends. How many stickers does each friend get?

Mathtastic Foundation Level
Sharing
Numbers 1-6

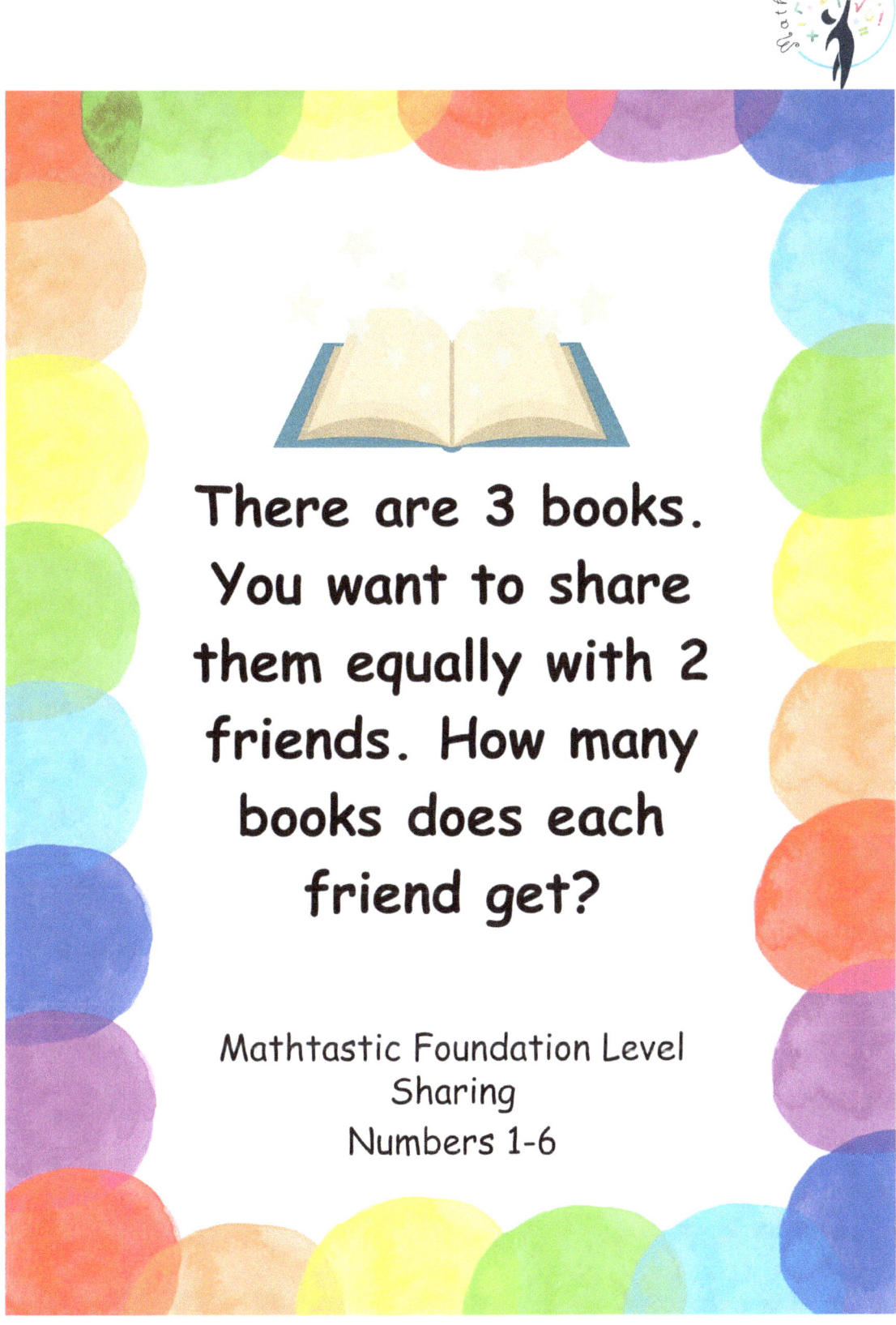

There are 3 books. You want to share them equally with 2 friends. How many books does each friend get?

Mathtastic Foundation Level
Sharing
Numbers 1-6

You have 4 blocks. You want to share them equally with 3 friends. How many blocks does each friend get?

Mathtastic Foundation Level
Sharing
Numbers 1-6

There are 2 frogs. You want to share them equally with 2 ponds. How many frogs does each pond get?

Mathtastic Foundation Level
Sharing
Numbers 1-6

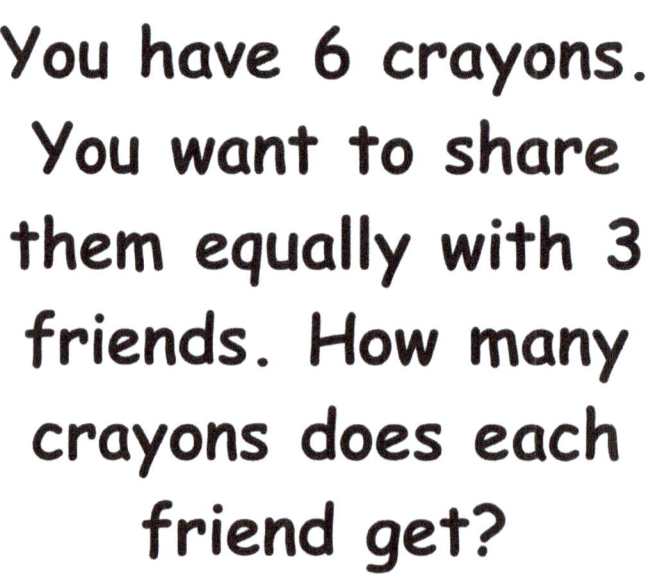

You have 6 crayons. You want to share them equally with 3 friends. How many crayons does each friend get?

Mathtastic Foundation Level
Sharing
Numbers 1-6

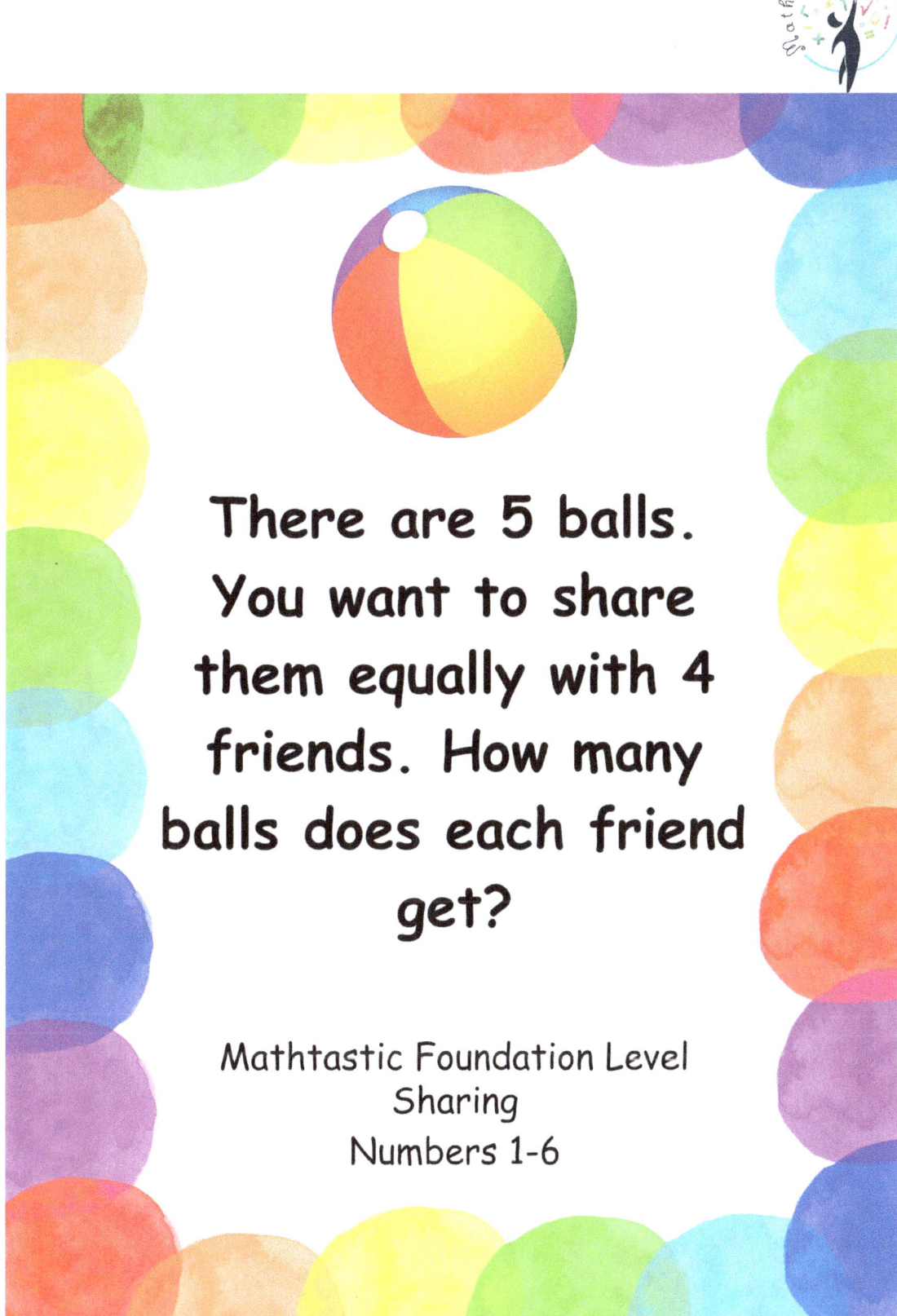

There are 5 balls. You want to share them equally with 4 friends. How many balls does each friend get?

Mathtastic Foundation Level
Sharing
Numbers 1-6

You have 3 stickers. You want to share them equally with 2 friends. How many stickers does each friend get?

Mathtastic Foundation Level
Sharing
Numbers 1-6

Find out more about Mathtastic at
www.mathtastic.com.au

Our other publications:

Level 1
Numbers 1-10

Level 2
Numbers 1-20

Level 3
Numbers 1-50

Coming next:

- Level 4 Numbers to 100
- Level 5 Numbers to 200
- Fractions
- Decimals
- Percentages
- Times Tables

www.ingramcontent.com/pod-product-compliance
Lightning Source LLC
Chambersburg PA
CBHW081418300426
44109CB00019BA/2337